PRINCIPLES OF

Total

Quality

Vincent K. Omachonu, Ph.D.
University of Miami
Coral Gables, Florida

Joel E. Ross, Ph.D.
Florida Atlantic University
Boca Raton, Florida

S_L^t

St. Lucie Press

Library of Congress Cataloging-in-Publication Data

Ross, Joel E.
    Principles of total quality / Joel E. Ross, Vincent K. Omachonu.
      p. cm.
    Includes bibliographical references and index.
    ISBN 0-9634030-6-0
    1. Total quality management.   2. Quality control.   3. Quality
control—Standards.   I. Omachonu, Vincent K. II. Title.
    HD62.15.R668   1994
    658.5'62—dc20   93-46439
                                              CIP

    Direct all inquiries to St. Lucie Press, Inc., 100 E. Linton Blvd., Suite 403B, Delray Beach, Florida 33483.

Phone: (407) 274-9906
Fax: (407) 274-9927

S^{t}_{L}

Published by
St. Lucie Press
100 E. Linton Blvd., Suite 403B
Delray Beach, FL 33483

PREFACE

No management issue since the scientific management movement pioneered by Frederick Taylor has had the impact of the current concern with quality and productivity. Quality expert J. M. Juran calls it a major phenomenon in this age.

It is generally agreed among U.S. executives that quality in products and services is essential in order to maintain and improve competitiveness in global markets. There is unequivocal evidence to indicate that for the individual firm, quality means greater market share and profitability. Therein lies a paradox. Despite the known benefits of quality, many firms do not perceive it as a route to greater success. Others are unwilling to take the necessary actions to overcome bureaucratic inertia. This fixation with structure and the status quo is unfortunate. It is in opposition to quality improvement as well as the quality of work life and the people dimension of management. Significant potential exists in all organizations for improving productivity and quality.

This book is not about the "hard" science of statistical quality control, although this topic is treated along with the other applied tools and techniques that are necessary for implementation of a quality program. The book is about total quality management (TQM) and about operationalizing a philosophy of management, a corporate vision, and a strategic differentiation. This is achieved by integrating all functions and activities for the purpose of continuous improvement and customer satisfaction.

The book is written for practitioners who are responsible for developing and implementing TQM programs in their organizations, whether in manufacturing or service industries. It is also an excellent text for college students as well as for organizational development and training programs.

The book is organized into four parts. In Part I, Management of Total Quality, the core ideas of TQM are provided. These are based on the Malcolm Baldrige Award and the criteria that are universally recognized as the standards for quality programs. These criteria are expanded and integrated with supporting management principles. Also included is the important topic of benchmarking, as well as a discussion of work teams and the other basics of organizing for quality.

Part II contains a treatment of the basic tools and techniques that are needed for data gathering and analysis related to process improvement. Part III provides criteria for quality programs that are based on the Baldrige Award and for ISO 9000 as this latter movement accelerates in the United States. Also included is a discussion of the emerging concept of reengineering.

Part IV contains three comprehensive case studies that permit the reader, the student, or the seminar participant to test his or her TQM knowledge and skills against three real-life organizations.

AUTHORS

Dr. Vincent K. Omachonu is Assistant Professor of Industrial Engineering at the University of Miami. He received his Ph.D. in Industrial Engineering from the Polytechnic Institute of New York, Brooklyn. He earned masters degrees in Operations Research from Columbia University, New York and in Industrial Engineering from the University of Miami, Florida. His B.S. degree is also in Industrial Engineering from the University of Miami.

Dr. Omachonu has been active in the Institute of Industrial Engineers (IIE) for several years and is a senior member of the Institute. He has served as both president and director of the Miami chapter. Dr. Omachonu is also a member of the Health Systems Society of IIE, American Hospital Association, Scientific Research Society of America (Sigma Xi), Institute of Management Services, London, and the Association for Health Services Research.

Dr. Omachonu's research interests include total quality and productivity management, the study of white-collar/knowledge work, production systems, facilities location and layout, work measurement and systems design, healthcare quality and productivity management, and technology management. He has been a consultant to a number of multinational corporations.

Dr. Omachonu has published several articles in national and technical journals, proceedings, and books. He is the author of a book entitled *Total Quality and Productivity Management in Healthcare Organizations,* which won the 1993 IIE Joint Publishers Book-of-the-Year Award. He has also served as a reviewer for a number of journals.

Dr. Joel Ross is Senior Professor of Management at Florida Atlantic University in Boca Raton, Florida. He graduated from Yale University and received his doctorate in business administration from George Washington University. He has been Chairman of Management and Director of the MBA Program. Prior to his academic career, Dr. Ross was a Commander in the U.S. Navy.

Dr. Ross is widely known as a platform speaker, seminar leader, consultant, and author. He has developed and conducted management development programs for over one hundred companies and organizations in the areas of general management, strategy, productivity, and quality. He has been an invited lecturer on management topics in Israel, South Africa, Venezuela, Panama, India, Ecuador, the Philippines, and Japan.

His articles have appeared in such journals as *Journal of Systems Management, Business Horizons, Long Range Planning, Industrial Management, Personnel, Management Accounting,* and *Academy of Management Review.* He is the author of thirteen books, including the landmark *Management Information Systems, People, Profits, and Productivity,* and *Total Quality Management,* which has been adopted by over 250 colleges and universities.

Dr. Ross has the reputation of being able to integrate academic principle with real-world practice.

IN MEMORIAM

W. Edwards Deming

1900–1993

The Guru of Quality

CONTENTS

PART III: CRITERIA for QUALITY PROGRAMS 269

PART IV: CASES in QUALITY 309

I

MANAGEMENT
of
TOTAL QUALITY

The concepts of quality and good management principles have been around for some time, but each has been treated separately and the two sometimes considered unrelated topics. Both concepts are integrated in Part I, where the idea is advanced that quality requires the continuing application of management principles.

In Chapter 1, the concept of total quality management (TQM) is introduced, the emergence of the movement is traced, and the pioneers who developed the principles and techniques are identified. In Chapter 2, the need for top management support and involvement is outlined, and how this should be reflected in the corporate culture and supporting management systems is described.

How information systems serve both strategic and operational needs and link organizational functions is described in Chapter 3. Elements of system design are addressed.

In Chapter 4, the process of strategy development is explained and the role of quality as the differentiating factor in strategy is explored. The idea of involvement and empowerment as the critical dimension of human resource management is presented in Chapter 5, and the need to make quality a central ingredient of these methods is examined.

The emergence of process control rather than final inspection as a means to continuous improvement is traced in Chapter 6. Quality function deployment and just-in-time are discussed. The measurement and improvement of customer satisfaction and standards for customer retention are covered in Chapter 7. In Chapter 8, the steps involved in benchmarking—comparing oneself to best-in-class organizations—are provided.

The systems approach to a TQM organization style and how to achieve cross-functional integration with teams are described in Chapter 9. Included in Chapter 10 are the basics of productivity management and how productivity is achieved through quality improvement.

The cost of quality is covered in Chapter 11, as well as how to measure the cost of not meeting customer requirements—the cost of doing things wrong. The use of quality cost information is also discussed.

1

TOTAL QUALITY MANAGEMENT and the REVIVAL of QUALITY in the UNITED STATES

Total quality management (TQM) is the integration of all functions and processes within an organization in order to achieve continuous improvement of the quality of goods and services. The goal is customer satisfaction.

■ Xerox, one of two 1989 winners of the Malcolm Baldrige National Quality Award, has come a long way since the 1970s. The company that invented the dry-paper copier saw its share of the North American market plunge from 93 to 40 percent. Japanese competition was selling copiers for less than the cost of manufacture at Xerox. By using TQM (known as Leadership through Quality at Xerox), the company has gained market share in all key markets worldwide and builds five of the six highest quality copiers in the world. The company has since learned to apply quality beyond the manufacturing confines in all the functions of the organization.

■ Motorola, another winner of the Malcolm Baldrige Award, has received more awards for excellence as a supplier than any other United States company and is widely acknowledged as a quality leader. This is quite an improvement over the early 1980s, when Chairman Bob Galvin was calling for a three-year, 20 percent surcharge on all imported manufactured goods in an attempt to counteract the threat from the Orient. The company applies TQM to every aspect of its operations and six sigma to every significant business process.

Of all the management issues faced in the last decade, none has had the impact of or caused as much concern as quality in American products and services. A report by the Conference Board indicates that senior executives in the United States agree that the banner of total quality is essential to ensure competitiveness in global markets. Quality expert J. M. Juran calls it a major phenomenon in this age.[1] This concern for quality is not misplaced.

The interest in quality is due, in part, to foreign competition and the trade deficit.[2] Analysts estimate that the vast majority of United States businesses will continue to face strong competition from the Pacific Rim and the European Economic Community for the remainder of the 1990s and beyond.[3] This comes in the face of a serious erosion of corporate America's ability to compete in global markets over the past 20 years.

The problem has not gone unnoticed by government officials, corporate executives, and the public at large. The concern of the President and Congress culminated in the enactment of the Malcolm Baldrige National Quality Improvement Act of 1987 (Public Law 100-107), which established an annual United States National Quality Award. The concern of business executives is reflected in their perceptions of quality. In a 1989 American Society for Quality Control (ASQC) survey, 54 percent of executives rated quality of service as extremely critical and 51 percent rated quality of product as extremely critical.[4] Seventy-four percent gave American-made products less than eight on a ten-point scale for quality. Similarly, a panel of Fortune 500 executives agreed that American products deserved no better than a grade of C+.

Public opinion regarding American-made products is somewhat less than enthusiastic. In a 1988 ASQC survey of consumer perceptions, less than one-half gave American products high marks for quality.[5] Employees also have misgivings about quality in general and, more specifically, about quality in the companies in which they work. They believe that there is a significant gap between what their companies say and what they do. More importantly, employees believe that their talents, abilities, and energies are not being fully utilized for quality improvement.[6]

Despite the pessimism reflected by these groups, progress is being made. In a 1991 survey of American owners of Japanese-made cars, 32 percent indicated that their next purchase will be a domestic model, and the reason given most often was the improved quality of cars built in the United States.[7] Ford's "Quality Is Job One" campaign may have been a contributing factor. There is also evidence that quality has become a competitive marketing strategy in the small business community, as Americans are beginning to shun mass-produced, poorly made, disposable products.

Other promising developments include the increasing acceptance of TQM as a philosophy of management and a way of company life. It is essential that this trend continue if American companies are to remain competitive in global markets. Customers are becoming more demanding and international competition more fierce. Companies that deliver quality will prosper in the next century.

The CONCEPT of TQM

TQM is based on a number of ideas. It means thinking about quality in terms of all functions of the enterprise and is a start-to-finish process that integrates interrelated functions at all levels. It is a systems approach that considers every interaction between the various elements of the organization. Thus, the overall effectiveness of the system is higher than the sum of the individual outputs from the subsystems. The subsystems include all the *organizational functions* in the life cycle of a product, such as (1) design, (2) planning, (3) production, (4) distribution, and (5) field service. The *management* subsystems also require integration, including (1) strategy with a customer focus, (2) the tools of quality, and (3) employee involvement (the linking process that integrates the whole). A corollary is that any product, process, or service can be improved, and a successful organization is one that consciously seeks and exploits opportunities for improvement at all levels. The load-bearing structure is customer satisfaction. The watchword is *continuous improvement.*

Following an international conference in May 1990, the Conference Board summarized the key issues and terminology related to TQM:

■ The **cost of quality** as the measure of non-quality (not meeting customer requirements) and a measure of how the quality process is progressing.

■ A **cultural change** that appreciates the primary need to meet customer requirements, implements a management philosophy that acknowledges this emphasis, encourages employee involvement, and embraces the ethic of continuous improvement.

■ **Enabling mechanisms of change,** including training and education, communication, recognition, management behavior, teamwork, and customer satisfaction programs.

■ **Implementing TQM** by defining the mission, identifying the output, identifying the customers, negotiating customer requirements, developing a "supplier specification" that details customer objectives, and determining the activities required to fulfill those objectives.

■ **Management behavior** that includes acting as role models, use of quality processes and tools, encouraging communication, sponsoring feedback activities, and fostering and providing a supporting environment.[8]

ANTECEDENTS of MODERN QUALITY MANAGEMENT

Quality control as we know it probably had its beginnings in the factory system that developed following the Industrial Revolution. Production methods at that time were rudimentary at best. Products were made from non-standardized materials using non-standardized methods. The result was products of varying quality. The only real standards used were measures of dimensions, weight, and in some instances purity. The most common form of quality control was inspection by the purchaser, under the common law rule of *caveat emptor*.[9]

Much later, around the turn of this century, Frederick Taylor developed his system of scientific management, which emphasized productivity at the expense of quality. Centralized inspection departments were organized to check for quality at the end of the production line. An extreme example of this approach was the Hawthorne Works at Western Electric Company, which at its peak in 1928 employed 40,000 people in the manufacturing plant, 5,200 of whom were in the inspection department. The control of quality focused on final inspection of the manufactured product, and a number of techniques were developed to enhance the inspection process. Most involved visual inspection or testing of the product following manufacture. Methods of statistical quality control and quality assurance were added later. Detecting manufacturing problems was the overriding focus. Top management moved away from the idea of managing to achieve quality and, furthermore, the work force had no stake in it. The concern was limited largely to the shop floor.

Traditional quality control measures were (and still are) designed as defense mechanisms to prevent failure or eliminate defects.[10] Accountants were taught (and are still taught) that expenditures for defect prevention were justified only if they were less than the cost of failure. Of course, cost of failure was rarely computed. (Cost of quality is discussed further in Chapter 11.)[11]

Following World War II, the quality of products produced in the United States declined as manufacturers tried to keep up with the demand for non-military goods that had not been produced during the war. It was during this period that a number of pioneers began to advance a methodology of quality control in manufacturing and to develop theories and practical techniques for improved quality. The most visible of these pioneers were W. Edwards Deming, Joseph M. Juran, Armand V. Feigenbaum, and Philip Crosby.[12] It was a great loss to the quality movement when Deming died in December 1993 at the age of 93.

The QUALITY GURUS

Deming, the best known of the "early" pioneers, is credited with popularizing quality control in Japan in the early 1950s. Today he is regarded as a national hero in that country and is the father of the world-famous Deming Prize for Quality. He is best known for developing a system of statistical quality control, although his contribution goes substantially beyond those techniques.[13] His philosophy begins with top management but maintains that a company must adopt the fourteen points of his system at all levels. He also believes that quality must be built into the product at all stages in order to achieve a high level of excellence. While it cannot be said that Deming is responsible for quality improvement in Japan or the United States, he has played a substantial role in increasing the visibility of the process and advancing an awareness of the need to improve.

Deming defines quality as *zero defects* or *reduced variations* and relies on statistical process control as the essential problem-solving technique for distinguishing systemic causes from special causes. The pursuit of quality results in lower costs, improved productivity, and competitive success. Although it is the worker who will ultimately produce quality products, Deming stresses worker pride and satisfaction rather than the establishment of quantifiable goals. His overall approach focuses on improvement of the process, in that the system, rather than the worker, is the cause of process variation.

Deming's *universal fourteen points* for management are summarized as follows:

1. Create consistency of purpose with a plan.
2. Adopt the new philosophy of quality.
3. Cease dependence on mass inspection.
4. End the practice of choosing suppliers based solely on price.
5. Identify problems and work continuously to improve the system.
6. Adopt modern methods of training on the job.
7. Change the focus from production numbers (quantity) to quality.
8. Drive out fear.
9. Break down barriers between departments.
10. Stop requesting improved productivity without providing methods to achieve it.
11. Eliminate work standards that prescribe numerical quotas.
12. Remove barriers to pride of workmanship.
13. Institute vigorous education and retraining.
14. Create a structure in top management that will emphasize the preceding thirteen points every day.

Juran, like Deming, was invited to Japan in 1954 by the Union of Japanese Scientists and Engineers (JUSE). His lectures introduced the managerial dimensions of planning, organizing, and controlling and focused on the responsibility of management to achieve quality and the need for setting goals.[14] Juran defines quality as *fitness for use* in terms of design, conformance, availability, safety, and field use. Thus, his concept more closely incorporates the point of view of the customer. He is prepared to measure everything and relies on systems and problem-solving techniques. Unlike Deming, he focuses on top-down management and technical methods rather than worker pride and satisfaction.

Juran's ten steps to quality improvement are

1. Build awareness of opportunities to improve.
2. Set goals for improvement.
3. Organize to reach goals.
4. Provide training.
5. Carry out projects to solve problems.
6. Report progress.
7. Give recognition.
8. Communicate results.
9. Keep score.
10. Maintain momentum by making annual improvement part of the regular systems and processes of the company.

Juran is the founder of the Juran Institute in Wilton, Connecticut. He promotes a concept known as Managing Business Process Quality, which is a technique for executing cross-functional quality improvement. Juran's contribution may, over the longer term, be greater than Deming's because Juran has the broader concept, while Deming's focus on statistical process control is more technically oriented.[15]

Armand Feigenbaum, like Deming and Juran, achieved visibility through his work with the Japanese. Unlike the latter two, he used a total quality control approach that may very well be the forerunner of today's TQM. He promoted a system for integrating efforts to develop, maintain, and improve quality by the various groups in an organization. To do otherwise, according to Feigenbaum, would be to inspect for and control quality after the fact rather than build it in at an earlier stage of the process.

Philip Crosby, author of the popular book *Quality is Free*,[16] may have achieved the greatest commercial success by promoting his views and founding the Quality College in Winter Park, Florida. He argues that poor quality in the average firm costs about 20 percent of revenues, most of which could be avoided by adopting good quality practices. His "absolutes" of quality are

- Quality is **defined** as conformance to requirements, not "goodness."
- The **system** for achieving quality is prevention, not appraisal.
- The performance **standard** is zero defects, not "that's close enough."
- The **measurement** of quality is the price of non-conformance, not indexes.[17]

Crosby stresses motivation and planning and does not dwell on statistical process control and the several problem-solving techniques of Deming and Juran. He states that quality is free because the small costs of prevention will always be lower than the costs of detection, correction, and failure. Like Deming, Crosby has his own *fourteen points:*

1. **Management commitment.** Top management must become convinced of the need for quality and must clearly communicate this to the entire company by written policy, stating that each person is expected to perform according to the requirement or cause the requirement to be officially changed to what the company and the customers really need.

2. **Quality improvement team.** Form a team composed of department heads to oversee improvements in their departments and in the company as a whole.

3. **Quality measurement.** Establish measurements appropriate to every activity in order to identify areas in need of improvement.

4. **Cost of quality.** Estimate the costs of quality in order to identify areas where improvements would be profitable.

5. **Quality awareness.** Raise quality awareness among employees. They must understand the importance of product conformance and the costs of non-conformance.

6. **Corrective action.** Take corrective action as a result of steps 3 and 4.

7. **Zero defects planning.** Form a committee to plan a program appropriate to the company and its culture.

8. **Supervisor training.** All levels of management must be trained in how to implement their part of the quality improvement program.

9. **Zero defects day.** Schedule a day to signal to employees that the company has a new standard.

10. **Goal setting.** Individuals must establish improvement goals for themselves and their groups.

11. **Error cause removal.** Employees should be encouraged to inform management of any problems that prevent them from performing error-free work.

12. **Recognition.** Give public, non-financial appreciation to those who meet their quality goals or perform outstandingly.

13. **Quality councils.** Composed of quality professionals and team chairpersons, quality councils should meet regularly to share experiences, problems, and ideas.

14. **Do it all over again.** Repeat steps 1 to 13 in order to emphasize the never-ending process of quality improvement.

All of these pioneers believe that management and the system, rather than the workers, are the cause of poor quality. These and other trailblazers have largely absorbed and synthesized each other's ideas, but generally speaking they belong to two schools of thought: those who focus on technical processes and tools and those who focus on the managerial dimensions.[18] Deming provides manufacturers with methods to measure the variations in a production process in order to determine the causes of poor quality. Juran emphasizes setting specific annual goals and establishing teams to work on them. Crosby stresses a program of zero defects. Feigenbaum teaches total quality control aimed at managing by applying statistical and engineering methods throughout the company.

Despite the differences among these experts, a number of common themes arise:

1. Inspection is never the answer to quality improvement, nor is "policing."

2. Involvement of and leadership by top management are essential to the necessary culture of commitment to quality.

3. A program for quality requires organization-wide efforts and long-term commitment, accompanied by the necessary investment in training.

4. Quality is first and schedules are secondary.

Admiration for Deming's contribution is not confined to Japan. At the Yale University commencement in May 1991, Deming was awarded an honorary degree. The citation read:

> W. Edwards Deming, '28 PhD, *consultant in statistical studies.* For the past four decades, you have been the champion of quality management. You have developed a theory of management, based on scientific and statistical principles in which people remain the least predictable and most important part. Your scholarly insights and your wisdom have revolutionized industry. Yale is proud to confer upon you the degree of Doctor of Laws.[19]

ACCELERATING USE OF TQM

The increased acceptance and use of TQM is the result of three major trends: (1) reaction to increasing domestic and global competition, (2) the pervasive need to integrate the several organizational functions for improvement of total output of the organization as well as the quality of output within each function, and (3) the acceptance of TQM in a variety of service industries.

Aside from existing competitive pressures from Japan and the Pacific Rim countries, American firms are faced with the prospect of increasing competition from members of the European Economic Community. This concern is justified by the very nature of manufacturing strategy among European firms, where quality has replaced technology as the primary consideration.

Basic to the concept of TQM is the notion that quality is essential in all functions of the business, not just manufacturing. This is justified by reason of organization synergism: the need to provide quality output to internal as well as external customers and the facilitation of a quality culture and value system throughout the organization. Companies that commit to the concept of TQM apply quality improvement techniques in almost every area of product development, manufacturing, distribution, administration, and customer service.[20] Nowhere is the philosophy of "customer is king" more prevalent than in TQM. Customers are both external (including channels) and internal (including staff functions) to the business.

The paradigm of TQM applies to all enterprises, both manufacturing and service, and many companies in manufacturing, service, and information industries have reaped the benefits. Industries as diverse as telecommunications, public utilities, and health care have applied the principles of TQM.

Government agencies and departments have also joined the movement, although private sector efforts have been considerably more effective.[21] According to a 1992 General Accounting Office (GAO) special report, 68 percent of the federal organizations and installations surveyed had some kind of TQM effort underway. Productivity and quality improvement programs are expected to be initiated in almost 700 federal programs in 1993.[22] Defense contractors will be particularly affected as the government moves toward requiring suppliers to adopt the TQM concept.[23] Oregon State University is the first among academic institutions to make a commitment to adopt the principles of TQM throughout the organization.[24]

The widespread adoption of one or more approaches or principles of TQM does not mean that results have met expectations. According to the GAO survey mentioned earlier, only 13 percent of government agency employees actively participate in the TQM efforts.[25] Human resource professionals report a strong interest in TQM issues in 1993, ranking employee involvement, customer service, and TQM as the top three key issues, yet research shows that initiatives taken by organizations are not receiving as much praise as they did a few years ago.[26]

QUALITY and BUSINESS PERFORMANCE

The relationship between quality, profitability, and market share has been studied in depth by the Strategic Planning Institute of Cambridge, Massachusetts. The conclusion, based on performance data of about 3000 strategic business units, is unequivocal:

> One factor above all others—quality—drives market share. And when superior quality and large market share are both present, profitability is virtually guaranteed.[27]

> There is no doubt that relative perceived quality and profitability are strongly related. Whether the profit measure is return on sales or return on investment, businesses with a superior product/service offering clearly outperform those with inferior quality.[28]

Even producers of commodity or near-commodity products seek and find ways to distinguish their products through cycle time, availability, or other quality attributes.[29] In addition to profitability and market share, quality drives growth. The linkages between these correlates of quality are shown in Figure 1-1.

Quality can also reduce costs. This reduction, in turn, provides an additional competitive edge. Note that Figure 1-1 includes two types of quality: customer-driven quality and conformance or internal specification quality. The latter relates to appropriate product specifications and service standards that lead to cost reduction. As will be discussed in Chapter 11, there is an inverse relationship between internal or conformance quality and costs, and thus the phrase coined by Crosby: "Quality Is Free."[30] As quality improves, so does cost, resulting in improved market share and hence profitability and growth. This, in turn, provides a means for further investment in such quality improve-

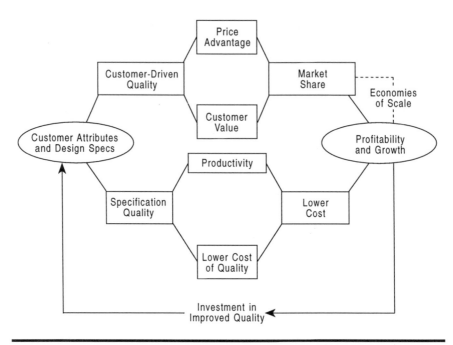

Figure 1-1   The Quality Circle

ment areas as research and development. The cycle goes on. In summary, improving both internal (conformance) quality and external (customer perceived) quality not only lowers cost of poor quality or "non-quality" but also serves as a driver for growth, market share, and profitability.

The rewards of higher quality are positive, substantial, and pervasive.[31] Findings indicate that attaining quality superiority produces the following organizational benefits:

1. Greater customer loyalty
2. Market share improvements
3. Higher stock prices
4. Reduced service calls
5. Higher prices
6. Greater productivity

SERVICE QUALITY VS. PRODUCT QUALITY

It is paradoxical that there is more concern in the United States for product quality than there is for quality of services and service industries. Despite the fact that only 21 percent of total employment in the United States is in industries that produce goods (excluding agriculture with approximately 3 percent),[32] the emphasis has historically been on manufacturing industries. Consider also that up to half of employment in manufacturing is in such staff or white-collar jobs as marketing, finance, or the many other activities not directly involved in physically producing products. If it is accepted that quality improvement can only be achieved through the actions of people, the conclusion emerges that possibly 90 percent or more of the potential for improvement lies in service industries and service jobs in manufacturing firms. The concept of "white-collar quality" is becoming increasingly recognized as the service sector grows.[33]

Despite this rather obvious need for quality service, people directly employed in manufacturing functions tend to focus on production first and quality second. "Get out the production" and "meet the schedule" are common cries on many shop floors. A study conducted by David Garvin of Harvard Business School revealed that U.S. supervisors believed that a deep concern for quality was lacking among workers and that quality as an objective in manufacturing was secondary to the primary goal of meeting production schedules. This same conclusion is suggested in the experiences of over 100 companies. Supervisors almost invariably set targets related to productivity and cost reduction rather than quality improvement.[34]

This seeming manufacturing-service paradox is unusual in view of the several considerations which suggest that the emphasis on services should be substantially increased. The first of these considerations is the "bottom line" factor. Studies have shown that companies rated highly by their customers in terms of service can charge close to 10 percent more than those rated poorly.[35] People will go out of their way and pay more for good service, which indicates the importance placed on service by customers. Conference Board reports concluded that the strongest complaints of customers were registered not for products but rather for services. Recognizing this, executives rate quality of service as a more critical issue than quality of product.[36] Tom Peters, co-author of *Search for Excellence,* scolds U.S. manufacturers for allowing quality to deteriorate into a mindless effort to copy the Japanese and suggests that the best approach is to learn from America's leading service companies.[37]

▪ Taking a cue from Domino's Pizza, their Michigan-based neighbor, Doctor's Hospital in Detroit is promising to see its emergency room patients in 20 minutes or the care will be free. During the first three weeks of the offer, no patients have been treated free of charge and the number of patients has been up 30 percent.[38]

As a strategic issue, customer service can be considered a major dimension of competitiveness. In the most exhaustive study in its history, the American Management Association surveyed over 3000 international respondents;[39] 78 percent identified improving quality and service to customers as *the* key to competitive success, and 92 percent indicated that providing superior service is one of their key responsibilities, regardless of position. To say that your competitive edge is price is to admit that your products and services are commodities.

After being viewed as a manufacturing problem for most of the past decade, quality has now become a service issue as well. TQM relates not only to the product, but to all the services that accompany it as well.

In many ways defining and controlling quality of service is more difficult than quality assurance of products. Unlike manufacturing, service industries share unique characteristics that make the process of quality control less manageable but no less important. Moreover, the level of quality expected is less predictable. Service company operations are affected by several characteristics, including the intangible nature of the output and the inability to store the output. Other distinguishing characteristics include:

1. Behavior of the delivery person
2. Image of the organization
3. The customer present during the production process and performing the final inspection
4. The measure of output is difficult to define
5. Variance and acceptance ranges may not apply
6. Adjusting the control system if the customer is present[40]

However, the most significant problem with the delivery of services is that it is typically measured at the customer interface—the one-on-

one, face-to-face interaction between supplier and customer. If a problem exists, it is already too late to fix it.[41]

Wall Street Journal, **March 4, 1993**

Vice President Gore's sphere expanded yesterday with the announcement that he will lead the latest White House effort to answer the call for change in Washington: a task force that will supposedly examine each federal agency for ways to cut spending and improve services. The "total quality management" effort is an idea borrowed from industry.

QUESTIONS for DISCUSSION

1-1 Give one or more examples of products made in Japan or Western Europe that are superior in quality to American-made products. How do you explain this difference?

1-2 Illustrate how the TQM concept can integrate design, engineering, manufacturing, and service.

1-3 Explain why quality should be better by following the TQM concept than in a system that depends on final inspection.

1-4 What common elements or principles can you identify among (1) the Baldrige criteria and (2) Deming, Juran, and Crosby?

1-5 Describe how increased market share and profitability might result from improved quality.

1-6 Select one staff department (e.g., accounting, finance, marketing services, human resources) and describe how this department can deliver quality service to its *internal* customers.

ENDNOTES

1. J. M. Juran, "Strategies for World Class Quality," *Quality Progress,* March 1991, p. 81.
2. Armand V. Feigenbaum, "America on the Threshold of Quality," *Quality,* Jan. 1990, p. 16. Feigenbaum, a pioneer and current expert in quality, estimates that

TQM could mean a 7 percent increase in the country's gross national product. See Armand V. Feigenbaum, "Quality: An International Imperative," *Journal for Quality and Participation,* March 1991, p. 16. Estimates from the United States Chamber of Commerce Department suggest that nearly 75 percent of all products manufactured in the United States are targets for strong competition from imports.

3. Ronald M. Fortuna, "The Quality Imperative," *Executive Excellence,* March 1990, p. 1. It is expected that competition from Europe may become more severe than that from Japan. Despite the surface congeniality at the G-7 meetings in London in mid-July 1991, it is apparent that the European Economic Community plans a united front against the United States. German Chancellor Helmut Kohl states, "Europe's return to its original unity means that the '90s will be the decade of Europe, not Japan. This is Europe's hour." (Peter Truell and Philip Revzin, "A New Era Is at Hand in Global Competition: U.S. vs. United Europe," *Wall Street Journal,* July 15, 1991, p. 1.)

4. American Society for Quality Control, *Quality: Executive Priority or Afterthought?* Milwaukee: ASQC, 1989.

5. American Society for Quality Control, *'88 Gallup Survey: Consumers' Perceptions Concerning the Quality of American Products and Services,* Milwaukee: ASQC, 1988, p. iv. This survey and the survey cited in Endnote 4 were conducted by the Gallup Organization.

6. American Society for Quality Control, *Quality: Everyone's Job, Many Vacancies,* Milwaukee: ASQC, 1990. This survey was conducted by the Gallup Organization. Other findings include: (1) where quality improvement programs exist the level of participation among employees is actually higher in small companies and service companies, (2) people who participate in quality improvement activities are more satisfied than non-participants with the rate of quality improvement their companies have been able to achieve.

7. A survey conducted by Integrated Automotive Resources, Wayne, Pennsylvania.

8. David Mercer, "Total Quality Management: Key Quality Issues," in *Global Perspectives on Total Quality,* New York: Conference Board, 1991, p. 11. See also Walter E. Breisch, "Employee Involvement," *Quality,* May 1990, pp. 49–51; John Hauser, "The House of Quality," *Harvard Business Review,* May/June 1988, pp. 63–73; W. F. Wheaton, "The Journey to Total Quality: A Fundamental Strategic Renewal," *Business Forum,* Spring 1989, pp. 4–7.

9. Claude S. George, Jr., *The History of Management Thought,* Englewood Cliffs, N.J.: Prentice-Hall, 1972, p. 53.

10. David A. Garvin, "Competing on the Eight Dimensions of Quality," *Harvard Business Review,* Nov./Dec. 1987, pp. 101–109. This same author has provided an excellent description of the background of quality developments in his book *Managing Quality* (New York: Free Press, 1988).

11. See Joel E. Ross and David E. Wegman, "Quality Management and the Role of the Accountant," *Industrial Management,* July/Aug. 1990, pp. 21–23.

12. Yunum Kathawala, "A Comparative Analysis of Selected Approaches to Quality," *International Journal of Quality and Reliability Management,* Vol. 6 Issue 5, 1989, pp. 7–17. Other writers and researchers with less visibility than those mentioned here have contributed to the literature. For a review of this rapidly expanding literature, see Jayant V. Saraph, P. George Benson, and Roger G. Schroeder, "An

Instrument for Measuring the Critical Factors of Quality Management," *Decision Sciences,* Vol. 20, 1989. The research described in this article identified 120 prescriptions for effective quality management, which were subsequently grouped into 8 categories that are quite similar to the Baldrige Award criteria: (1) the role of management leadership and quality policy, (2) the role of the quality department, (3) training, (4) product/service design, (5) supplier quality management, (6) process management, (7) quality data and reporting, and (8) employee relations.

13. W. Edwards Deming, *Quality, Productivity, and Competitive Position,* Cambridge, Mass.: Center for Advanced Engineering Study, Massachusetts Institute of Technology, 1982. See also W. Edwards Deming, *Out of the Crisis,* Cambridge, Mass.: Center for Advanced Engineering Study, Massachusetts Institute of Technology, 1982.

14. Juran's early approach appears in J. M. Juran, *Quality Control Handbook,* New York: McGraw-Hill, 1951. For more recent contributions, see (all by Juran) "The Quality Trilogy," *Quality Progress,* Aug. 1986, pp. 19–24; "Universal Approach to Managing Quality," *Executive Excellence,* May 1989, pp. 15–17; "Made in USA—A Quality Resurgence," *Journal for Quality and Participation,* March 1991, pp. 6–8; "Strategies for World Progress," *Quality Progress,* March 1991, pp. 81–85.

15. "Dueling Pioneers," *Business Week,* Oct. 25, 1991. This is a special report and bonus issue of *Business Week* entitled "The Quality Imperative."

16. Philip Crosby, *Quality Is Free,* New York: McGraw-Hill, 1979.

17. From *Quality,* a promotional brochure by Philip Crosby Associates, Inc.

18. Sara Jackson, "Calling in the Gurus," *Director (UK),* Oct. 1990, pp. 95–101. In this article, the author reports that in the U.K., it is not the quality gurus as much as government initiatives that have been responsible for raising quality awareness.

19. Marc Wortman, "Commencement," *Yale Alumni Magazine,* Summer 1991, p. 61. One of the authors happens to be a Yale graduate, although not of the class of '28.

20. Daniel M. Stowell, "Quality in the Marketing Process," *Quality Progress,* Oct. 1989, pp. 57–62. For the sales function, see Walt Williams, "Quality: An Old Objective but a New Strategy." For R & D, see Michael F. Wolff, "Quality in R & D—It Starts With You," *Marketing News,* Oct. 15, 1990, pp. 16–22.

21. Stanley Blacker, "Data Quality and the Environment," *Quality,* April 1990, pp. 38–42.

22. Carolyn Burstein and Kathleen Sediak, "The Federal Quality and Productivity Improvement Effort," *Quality Progress,* Oct. 1988, pp. 38–41.

23. General Dynamics and McDonnell Douglas are among the defense contractors that have achieved improvement through TQM. See Bruce Smith and William B. Scott, "Douglas Tightens Controls to Improve Performance," *Aviation Week & Space Technology,* Jan. 4, 1990, pp. 16–20. See also Glenn E. Hayes, "Three Views of TQM," *Quality,* April 1990, pp. 19–24.

24. Edwin L. Coate, "TQM at Oregon State University," *Journal for Quality and Participation,* Dec. 1990, pp. 90–101.

25. Jennifer Jordan, "Everything You Wanted to Know About TQM," *Public Manager,* Winter 1992–1993, pp. 45–48.

26. Karen Matthes, "A Look Ahead for '93," *HR Focus,* Jan. 1993, pp. 1, 4. See also

Richard Y. Chang, "When TQM Goes Nowhere," *Training and Development,* Jan. 1993, pp. 22–29.

27. Robert D. Buzzell and Bradley T. Gale, *The PIMS Principles: Linking Strategy to Performance,* New York: The Free Press, 1987, p. 87. PIMS is the acronym for Profit Impact of Market Strategy. A PIMS study in Canada reached a similar conclusion. See William Band, "Quality Is King for Marketers," *Sales and Marketing Management in Canada,* March 1989.

28. Robert D. Buzzell and Bradley T. Gale, *The PIMS Principles: Linking Strategy to Performance,* New York: The Free Press, 1987, p. 107.

29. A good example is the "Perdue Chicken" produced by Perdue Farms. Owner Frank Perdue set out to differentiate his chicken by color, freshness, availability, and meat-to-bone ratio. These criteria of quality, as defined by the customer, led the company to growth and improved market share and profitability. See Diane Feldman, "Building a Better Bird," *Management Review,* May 1989, pp. 10–14.

30. Tom Peters reported in *Thriving on Chaos* (New York: Knopf, 1987) that experts agree that poor quality can cost about 25 percent of the people and assets in a manufacturing firm and up to 40 percent in a service firm.

31. See Joel E. Ross and David Georgoff, "A Survey of Productivity and Quality Issues in Manufacturing: The State of the Industry," *Industrial Management,* Jan./Feb. 1991.

32. United States Bureau of Labor Statistics, *Monthly Labor Review,* Nov. 1989. Reported in *U.S. Statistical Abstract,* 1990, p. 395.

33. For example, Campbell USA has targeted the administrative and marketing side of the corporation in its latest quality program, "Quality Proud." See Herbert M. Baum, "White-Collar Quality Comes of Age," *Journal of Business Strategy,* March/April 1990, pp. 34–37.

34. David Garvin, "Quality Problems, Policies, and Attitudes in the United States and Japan: An Exploratory Study," *Academy of Management Journal,* Dec. 1986, pp. 653–673.

35. Frank K. Sonnenberg, "Service Quality: Forethought, Not Afterthought," *Journal of Business Strategy,* Sep./Oct. 1989, pp. 56–57.

36. American Society for Quality Control, *Quality: Executive Priority or Afterthought?"* Milwaukee: ASQC, 1989, p. 8. In this survey conducted by the Gallup Organization, 57 percent of service company executives rated service quality as extremely critical (10 on a scale of 1 to 10), while only 50 percent of industrial company executives gave service quality the same rating.

37. Tom Peters, "Total Quality Leadership. Let's Get It Right," *Journal for Quality and Participation,* March 1991, pp. 10–15.

38. "Hospital Delivers: Emergency Room Guarantees Care in 20 Minutes," Associated Press, July 15, 1991.

39. Reported in Eric R. Greenberg, "Customer Service: The Key to Competitiveness," *Management Review,* Dec. 1990, p. 29. Reported fully in AMA Research Report, *The New Competitive Edge,* New York: American Management Association, 1991.

40. See Terrence J. Smith, "Measuring a Customer Service Culture," *Retail Control,* Oct. 1989, pp. 15–18. See also Behshid Farsad and Ahmad K. Eishennawy, "Defining Service Quality Is Difficult for Service and Manufacturing Firms," *Industrial Engineering,* March 1989, pp. 17–19; Christian Gronroos, "Service Quality: The Six

Criteria of Good Perceived Service Quality," *Review of Business,* Winter 1989, pp. 10–13; Carol King, "Service Quality Assurance Is Different," *Quality Progress,* June 1985, pp. 14–18.

41. Lawrence Holpp, "Ten Steps to Total Service Quality," *Journal for Quality and Participation,* March 1990, pp. 92–96. The major steps referred to in the title include: (1) creating an awareness and a philosophy of constant improvement, (2) making the vision of the organization a personal vision for every employee, (3) empowering employees to act, (4) surveying customers personally, (5) measuring meaningful information, and (6) adopting a performance management system that rewards teamwork, improvement, and new behaviors consistent with interdepartmental cooperation.

2

LEADERSHIP

Getting quality results is not a short-term, instant-pudding way to improve competitiveness; implementing total quality management requires hands-on, continuous leadership.

Armand V. Feigenbaum

The story is told of three executives traveling on the same flight to an international conference. One executive was British, one Japanese, and one American. They were hijacked by terrorists and immediately before execution were offered an opportunity to make a last request. The Englishman asked to sing a verse of "God Save the Queen." The Japanese executive wanted to give a lecture on Japanese management. Upon hearing this, the American said: "Let me be the first one to be shot. I simply can't take another lecture on Japanese management."

The point of this story is that many U.S. managers are growing weary of such comparisons in which Americans appear to be second best. One such comparison involved a visit to several Japanese companies by seven Leadership Forum executives. The experience left them with a profound belief that the reason why Japanese companies are beating U.S. companies has little to do with trade barriers, culture, cost of capital, sympathetic unions, or a supportive government. They found that the primary reason is simply that the United States is being out-led and out-managed. With some notable exceptions, U.S. firms are lagging behind because they lack clear, consistent, and persistent leadership from the top. Joseph Jaworski, chairman of the American Leadership Forum, is among the many CEOs who suggest that quality depends upon a vision of excellence and that a vision becomes reality through excellent, compelling leadership.[1]

Some principles and practices of total quality management (TQM) may differ among firms and industries, but there is unanimous agreement as to the importance of leadership by top management in implementing TQM. Such leadership is a prerequisite to all strategy and action plans. According to Juran, it cannot be delegated.[2] Those firms that have succeeded in making total quality work for them have been able to do so because of strong leadership.[3] A U.S. General Accounting Office study concluded, "Ultimately, strong visionary leaders are the most important element of a quality management approach."[4]

Dr. Curt Reimann, Director of the Malcolm Baldrige National Quality Award, has reviewed hundreds of applications, including those of the award winners. His review of key excellence indicators of quality management is insightful and helpful for an award applicant or anyone using the Baldrige criteria as a benchmark to evaluate the quality of management. He summarizes the characteristics of excellent leadership as follows:[5]

Visible, committed, and knowledgeable. They promote the emphasis on quality and know the details and how well the company is doing. Personal involvement in education, training and recognition. Accessible to and routine contact with employees, customers and suppliers.

A missionary zeal. The leaders are trying to effect as much change as possible through their suppliers, through the government and through any other vehicle that promotes quality in the United States. Active in promotion of quality outside the company.

Aggressive targets. Going beyond incremental improvements and looking at the possibility of making large gains, getting the whole work force thinking about different processes—not just improving processes.

Strong drivers. Cycle time, zero defects, six sigma or other targets to drive improvements. Clearly defined customer satisfaction and quality improvement objectives.

Communication of values. Effecting cultural change related to quality. Written policy, mission, guidelines and other documented statements of quality values, or other bases for clear and consistent communications.

Organization. Flat structures that allow more authority at lower levels. Empowering employees. Managers as coaches rather than bosses. Cross-functional management processes and focus on internal as well as external customers. Interdepartmental improvement teams.

Customer contact. CEO and all senior managers are accessible to customers.

Two of the many companies that have received a great deal of visibility for their TQM programs are Westinghouse and IBM, both with divisions that have won the Baldrige Award. Westinghouse committed significant capital resources to support the quality improvement efforts of all Westinghouse divisions, including the creation of the first corporate-sponsored Productivity and Quality Center in the United States. The company's Total Quality Model (Figure 2-1) was developed for use by all division managers. Note that it is built upon a foundation of management leadership. The framework of IBM's corporate-wide quality program, called "Market Driven Quality" is shown in Figure 2-2. Again, note that the input or "driver" of the system is leadership.

David Kearns, Chairman and CEO of Xerox, explains how the company's "Leadership through Quality" process achieves commitment at every level: "training begins with our top-tier family work group—

Figure 2-1 The Westinghouse Total Quality Model

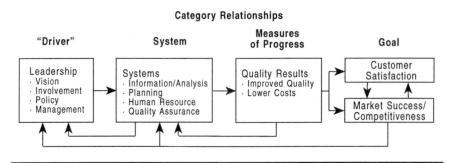

Figure 2-2 Framework of IBM's Market Driven Quality Program

my direct reports and me. It then cascades through the organizations led by senior staff, gradually spreading worldwide to some 100,000 employees.[6] This "cascading" reflects the leverage effect of good leadership at all levels. As one executive remarked, "it goes up, down and across the organization chart."

ATTITUDE and INVOLVEMENT of TOP MANAGEMENT

It would not be unfair to say that there has been a tendency among U.S. managers to focus on technology and hard assets rather than soft assets such as human resources and organizational competence.[7] The tendency has been to emphasize the organizational chart and the key control points within it. Many managers place priority on the budget and the business plan (to many these are the same) and assume that rational people will get on board and perform according to standard. This popular perception does not fit with leadership and a philosophy of quality.

It is axiomatic that organizations do not achieve quality objectives; people do. If there is a big push for quality or a new program, each employee is justifiably skeptical (the BOHICA syndrome—bend over, here it comes again). According to A. Blanton Godfrey, chairman and CEO of the Juran Institute, top management should be prepared to answer the specific question that may be posed by each member of the organization: "What do you want me to do tomorrow that is different from what I am doing today?"[8] Thus, top managers need to be ambidextrous. They must balance the need for the *structural* dimension (e.g., hierarchy, budgets, plans, controls, procedures) on the one hand with the *behavioral* or personnel dimension on the other. The two dimensions need not be in conflict.

> ▪ At 3M Company the leadership climate is proactive rather than reactive, externally focused rather than internally focused, and the quality perception views the totality of the business rather than just one aspect of it. In order to identify the gaps between its existing position and its vision of the future, 3M has developed "Quality Vision 2000" and implemented it through a process called Q90s which involves the total management system, making the process broader and deeper across the company worldwide.[9]

The commitment and involvement of management need to be demonstrated and visible. Speaking about his military experience, Dwight Eisenhower said: "They never listened to what I said, they always watch what I do."

Many managers send mixed signals. They endorse quality but reward bottom line or production. They insist on cost reduction even if it means canceling quality training. Still worse, some executives perceive the workers to be the cause of their quality problems.[10] This is hardly behavior that encourages individual involvement in decision making and personal "ownership" of the improvement process. Employee buy-in is unlikely in such a climate, where worker empowerment is talked about but not operationalized.

COMMUNICATION

Communication is inextricably linked in the quality process, yet some executives find it difficult to tell others about the plan in a way that will be understood. An additional difficulty is filtering. As top management's vision of quality gets filtered down through the ranks, the vision and the plan can lose both clarity and momentum. Thus, top management as well as managers and supervisors at all levels serve as translators and executors of top management's directive. The ability to communicate is a valuable skill at all levels, from front-line supervisor to CEO.

Quality-conscious companies are interested in the cost of poor communication in terms of both employee productivity and customer perception of product and service quality. More important than what is written or said is the recipient's perception of the message. Limited or inaccurate facts parceled out to employees may demoralize workers and lead to rumors.[11]

According to Peter Drucker, a true guru of management thought and practice, "The communications gap within institutions and between groups in society has been widening steadily—to a point where it threatens to become an unbridgeable gulf of total misunderstanding."[12] Having said that, he provides an easily understood and simple approach to help communicate the strategy, vision, and action plans related to TQM.

Communication is defined as the *exchange of information and understanding* between two or more persons or groups. Note the emphasis on exchange and understanding. Without understanding between sender and receiver concerning the message, there is no communication. The simple model is as follows:

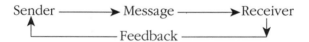

Unless sender gets feedback that receiver understands the message, no communication takes place. Yet most of us send messages with no feedback to indicate that the recipient (or percipient) has understood the message.

Despite the sorry state of communication, Drucker concludes that we do know something about communication in organizations and calls it "managerial communications." Communication is an extremely complex process. Many universities provide a doctoral program in the topic. At the risk of oversimplifying both communication theory and Drucker's approach, the essence of his principles can be paraphrased:

■ One can only communicate in terms of the recipient's language and perception, and therefore the message must be in terms of individual experience and perception. If the employee's perception of quality is "do a better job" or "keep the customer happy," it is unlikely that the message of TQM will be understood. Measures of quality are needed to ensure agreement on the meaning of the message.

■ Only the recipient can communicate—the communicator cannot. Thus, management systems (including training) should be designed from the point of view of the recipient and with a built-in mechanism for feedback. Feedback and thus the exchange of information should be based on some measure, target, benchmark, or standard.

■ All information is encoded, and prior agreement must be reached on the meaning of the code. Quality must be carefully defined and measures agreed upon.

■ Communication downward cannot work because it focuses on what we want to say. Communication should be upward.

■ Employees should be encouraged to set measurable goals.

Larry Appley, chairman emeritus of the American Management Association, has developed a company-wide productivity improvement program that has the model in Figure 2-3 as a centerpiece. Note that the direction of communication is *upward*. Recipient (subordinate) becomes sender, and sender (boss) becomes recipient. The message is specific and measurable, and the subordinate has ownership because he or she originated the message. Both parties can henceforth communicate about a message on which there is prior agreement. The Appley approach is therefore consistent with Drucker's ideas[13] and sound principles of communication. A modification tailored for a specific firm may be used as a vehicle for TQM implementation.

These concepts of effective communication can provide a practical approach for communicating about quality in the organization. It only

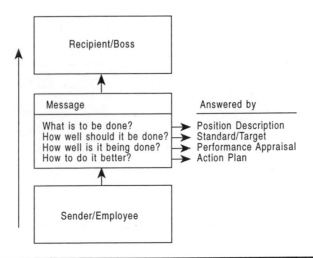

Figure 2-3 Effective Communication

remains to encode the message(s) in terms of recipient understanding. The vehicles for communicating about quality are selected components of the TQM system:

1. Training and development for both managers and employees. Managers must understand the processes they manage as well as the basic concept of systems optimization. Employee training should focus on the integration and appropriate use of statistical tools and problem-solving methods.

2. Participation at all levels in establishing benchmarks and measures of process quality. Involvement is both vertical in the hierarchy as well as horizontal by cross-functional teams.

3. Empowerment of employees by delegating authority to make decisions regarding process improvement within individual areas of responsibility, so that the individual "owns" the particular process step.

4. Quality assurance in all organization processes, not only in manufacturing or operations but in business and supporting processes as well. The objective throughout is continuous improvement.

5. Human resource management systems that facilitate contributions at all levels (up and down and across) the organizational chart.

The Digital Switching and Customer Service Division of Northern Telecom Canada Ltd. has received awards and international recognition for its quality systems and procedures. Continually communicating the importance of quality to its 5000 employees is considered vital by division management. Three internal communications specialists generate daily newsletters, monthly newspapers, and videos.[14] One method used by Westinghouse Electric Corporation to spread the word about quality to its 118,000 employees is an annual symposium. For two days each October, more than 600 employees gather to hear colleagues' TQM success stories. The goals for the symposium are for the chairman and senior management to energize employees and to provide attendees an opportunity to talk to each other.[15]

CULTURE

Culture is the pattern of shared beliefs and values that provides the members of an organization rules of behavior or accepted norms for conducting operations. It is the philosophies, ideologies, values, assumptions, beliefs, expectations, attitudes, and norms that knit an organization together and are shared by employees.[16]

For example, IBM's basic beliefs are (1) respect for the individual, (2) best customer service, and (3) pursuit of excellence. In turn, these beliefs are operationalized in terms of strategy and customer values. In simpler terms, culture provides a framework to explain "the way things are done around here."

Other examples of basic beliefs include:

Company	Basic belief
Ford	Quality is job one
Delta	A family feeling
3M	Product innovation
Lincoln Electric	Wages proportionate to productivity
Caterpillar	Strong dealer support; 24-hour spare parts support around the world
McDonald's	Fast service, consistent quality

Institutionalizing strategy requires a culture that supports the strategy. For most organizations a strategy based on TQM requires a significant if not sweeping change in the way people think. Jack Welch, head of General Electric and one of the most controversial and respected executives in America, states that cultural change must be sweeping—not incremental change but "quantum." His cultural transformation at GE calls for a "boundary-less" company where internal divisions blur, everyone works as a team, and both suppliers and customers are partners. His cultural concept of change may differ from Juran, who says that, "When it comes to quality, there is no such thing as improvement in general. Any improvement is going to come about project by project and no other way."[17] The acknowledged experts agree on the need for a cultural or value system transformation:

■ Deming calls for a transformation of the American management style.[18]

■ Feigenbaum suggests a pervasive improvement throughout the organization.[19]

■ According to Crosby, "Quality is the result of a carefully constructed culture, it has to be the fabric of the organization."[20]

It is not surprising that many executives hold the same opinions. In a Gallup Organization survey of 615 business executives, 43 percent rated a change in corporate culture as an integral part of improving quality. The needed change may be given different names in different companies. Robert Crandall, CEO of American Airlines, calls it an innovative environment,[21] while at DuPont it is "The Way People Think"[22] and at Allied Signal "Workers attitudes had to change."[23] Xerox specified a 5-year cultural change strategy called Leadership through Quality.[24] Tom Peters even adds what he calls "the dazzle factor."[25]

Successful organizations have a central core culture around which the rest of the company revolves. It is important for the organization to have a sound basis of core values into which management and other employees will be drawn. Without this central core, the energy of members of the organization will dissipate as they develop plans, make decisions, communicate, and carry on operations without a fundamental criteria of relevance to guide them. This is particularly true in decisions related to quality. Research has shown that quality means different things to different people and levels in the organization. Employees tend to think like their peers and think differently from those at other levels. This suggests that organizations will have considerable difficulty in improving quality unless core values are embedded in the organization.[26]

Commitment to quality as a core value for planning, organizing, and control will be doubly difficult when a concern for the practice is lacking. Research has shown that many U.S. supervisors believe that a concern for quality is lacking among workers and managers.[27] Where this is the case, the perceptions of these supervisors may become a self-fulfilling prophecy.

Embedding a Culture of Quality

It is one thing for top management to state a commitment to quality but quite another for this commitment to be accepted or embedded in the company. The basic vehicle for embedding an organizational culture is a teaching process in which desired behaviors and activities are learned through experiences, symbols, and explicit behavior. Once again, the components of the total quality system provide the vehicles for change. These components as well as other mechanisms of cultural change are summarized in Table 2-1. Above all, demonstration of commitment by top management is essential. This commitment is

Table 2-1 Cultural Change Mechanisms

Focus	From traditional	To quality
Plan	Short-range budgets	Future strategic issues
Organize	Hierarchy—chain of command	Participation/empowerment
Control	Variance reporting	Quality measures and information for self-control
Communication	Top down	Top down and bottom up
Decisions	Ad hoc/crisis management	Planned change
Functional management	Parochial, competitive	Cross-functions, integrative
Quality management	Fixing/one-shot manufacturing	Preventive/continuous, all functions and processes

demonstrated by behaviors and activities that are exhibited throughout the company. Categories of behaviors include:

- **Signaling.** Making statements or taking actions that support the vision of quality, such as mission statements, creeds, or charters directed toward customer satisfaction. Publix supermarkets' "Where shopping is a pleasure" and JC Penney's "The customer is always right" are examples of such statements.
- **Focus.** Every employee must know the mission, his or her part in it, and what has to be done to achieve it. What management pays attention to and how they react to crisis is indicative of this focus. When all functions and systems are aligned and when practice supports the culture, everyone is more likely to support the vision. Johnson and Johnson's cool reaction to the Tylenol scare is such an example.
- **Employee policies.** These may be the clearest expression of culture, at least from the viewpoint of the employee. A culture of quality can be easily demonstrated in such policies as the reward and promotion system, status symbols, and other human resource actions.

Executives at all levels could learn a lesson from David T. Kearns, chairman and chief executive officer of Xerox Corporation. In an article for the academic journal *Academy of Management Executive,* he describes the change at Xerox: "At the time Leadership-Through-Quality

Total Quality Transition

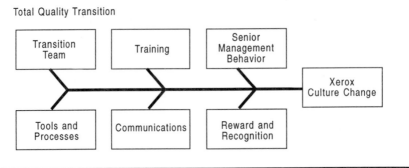

Figure 2-4 Transition to a Quality Culture at Xerox

was introduced, I told our employees that customer satisfaction would be our top priority and that it would change the culture of the company. We redefined quality as meeting the requirements of our customers. It may have been the most significant strategy Xerox ever embarked on."[28]

Among the changes brought about by the cultural change were the management style and the role of first-line management. Kearns continues: "We altered the role of first-line management from that of the traditional, dictatorial foreman to that of a supervisor functioning primarily as a coach and expediter."

Using a modification of the Ishikawa (fishbone) diagram, Xerox demonstrated (Figure 2-4) how the major component of the company's quality system was used for the transition to TQM.

MANAGEMENT SYSTEMS

No matter how comprehensive or lofty a quality strategy may be, it is not complete until it is put into action. It is only rhetoric until it has been implemented. Quality management systems are vehicles for change and should be designed to integrate all areas, not only the quality assurance department. They must be expanded throughout the company to include white-collar activities ranging from market research to shipping and customer service. They are directed toward achievement and commitment to purpose through four universal processes: (1) the specialization of task responsibilities through structure, (2) the provision of information systems that enable employees to know what they need to do in order to achieve goals, (3) the necessary achievement of results through action plans and projects, and (4) control through the establishment of benchmarks, standards, and feedback.

Each of these subsystems is the subject of a separate chapter in this book, but the implementation of each can only proceed from a base of clearly established goals. It is the specific task of top management to ensure that these goals are defined, disseminated, and implemented. Objectives in the areas of quality and productivity must be operationalized by establishing specific subobjectives for each function, department, or activity. Only then can courses of action be selected and plans implemented.

The problem, or conversely the opportunity, is to identify those *key* objectives and activities that are necessary in order to achieve a given strategy—in this case *quality*. The number of activities and processes in the typical organization is so large that a start-up quality management program cannot address all of them in the initial stages. Ultimately, every activity should be analyzed, its output evaluated in terms of value to both external and internal customers, and quality measures established.[29] Notwithstanding this longer term need, it is desirable to begin by setting goals only for those activities that are critical to achieving the mission statement and strategy.

What are these activities and processes that are critical to the mission of quality? The answer lies in identifying the key success factors that must be well managed if the mission or objective is to be achieved; that is, the limited number of areas in which results, if satisfactory, will ensure successful competitive performance for the organization.[30] Each activity or process can then be rated as to its importance. Advertising is a key success factor for Coca-Cola but not for McDonnell Douglas; design is critical to a hi-tech electronics firm but not to a bank.

This process can be used for any major objective, but it is also useful for providing a clear picture of things that must be done to implement a successful TQM program. Identification of key success factors emerges from three dimensions: (1) the drivers of quality such as cycle time reduction, zero defects, or six sigma; (2) operations that provide opportunities for reducing cost or improving productivity; and (3) the market side of quality, which relates to the salability of goods and services. These are converted to specific goals and targets which form the basis for subsequent programs and the universal processes identified earlier. Some U.S. managers have adopted ideas and language from Japanese companies, many of whom call the process *policy deployment.*[31]

CONTROL

The classical control process will require significant change if TQM is to be successful. Traditionally, control systems have been directed to the end use of preparation of financial statements. Focus has been on the components of the profit-and-loss statement. Quality control has historically followed a three-step process consisting of (1) setting standards, (2) reporting variances, and (3) correcting deviations. One source has defined control as "to review, to verify, to compare with standards, to use authority to bring about compliance, and to restrain."[32] In an organization that perceives control systems in this way, there is the danger that the system will become the end rather than the means. This is not to say that classical control does not have a place in quality management.

If a company is in a declining industry and its generic strategy (see Chapter 4) is low cost, or if its product or service is a near-commodity for which differentiation by quality is difficult, then the management system should be directed toward tight cost control, frequent detailed control reports, structured organization and responsibility, and incentives based on meeting strict quantitative targets.[33] If, on the other hand, a company has chosen TQM as a strategy and culture, significant changes in traditional control may be needed. The central idea is to meet the needs of people so that they can be productive. These needs are both personal and job related, and a system of control should be based on both. If employees "buy in" to quality, the control system should not be perceived as domination, but rather as a means toward self-control. The danger of classical control has been summarized by Peter Drucker:

> A system of controls which is not in conformity with this true, this only effective, this ultimate control of the organization which lies in its people decisions will therefore at best be ineffectual. At worst it will cause never-ending conflict and push the organization out of control.[34]

The difference between TQM control and traditional control is the difference between self-control and control by variance report, between continuous control and historical control, between feedforward and feedback.

Consider historical feedback control, as depicted in Figure 2-5. Assuming that there is a measure of output (a doubtful assumption for

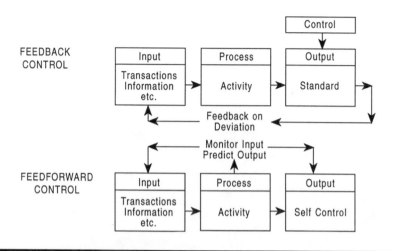

Figure 2-5 Feedback and Feedforward Control

most activities), the standard is compared to output, and variances are reported *after the fact*. The deviation has occurred and no amount of effort can change it. Typically, each period the first-line supervisor receives an after-the-fact statement of the quality control results for the entire plant. The worker receives nothing. This feedback is historical control by the numbers.

TQM control should be feedforward and predictive. Instead of measuring output after the fact, input is monitored by the individual or activity concerned, and output is forecast. If a deviation is predicted, action is taken to return to standard *before* the deviation occurs. There is no deviation because action is taken to avoid it before the fact. The concept is depicted in the bottom portion of Figure 2-5. The notion is fundamental for process control and continuous improvement of processes.

QUESTIONS for DISCUSSION

2-1 List the characteristics of excellent leadership for TQM.

2-2 Describe how leadership by top management is the *driver* of quality.

2-3 How can top management communicate the need for quality throughout the organization?

2-4 Describe how setting targets for quality improvement helps to establish a culture and climate.

2-5 Give an example of a company culture as reflected in a statement of basic beliefs. Would such a statement help to institutionalize a quality culture? If so, how?

2-6 How would an organization's commitment to quality facilitate or improve the following:

- The planning process
- Organization
- Control

2-7 Choose a manufacturing company and a service company. Identify a key activity for improving quality in each.

ENDNOTES

1. See Richard C. Whiteley, "Creating a Customer Focus," *Executive Excellence,* Sep. 1990 pp. 9–10.
2. J. M. Juran, "Made in USA—A Quality Resurgence," *Journal for Quality and Participation,* March 1991 pp. 6–8.
3. Thomas C. Gibson, "Helping Leaders Accept Leadership of Total Quality Management," *Quality Progress,* Nov. 1990, pp. 45–47.
4. U.S. General Accounting Office, *Quality Management Scoping Study,* Washington, D.C.: U.S. General Accounting Office, Dec. 1990, p. 25.
5. Curt W. Reimann, "Winning Strategies for Quality Improvement," *Business America,* March 25, 1991, pp. 8–11.
6. David T. Kearns, "Leadership through Quality," *Academy of Management Executive,* Vol. 4 No. 2, 1990, p. 87.
7. Michael Moacoby, "How to Be a Quality Leader," *Research-Technology Management,* Sep./Oct. 1990, pp. 51–52. See also Brian L. Joiner and Peter R. Scholtes, "The Quality Manager's New Job," *Quality Progress,* Oct. 1986, pp. 52–56.
8. A. Blanton Godfrey, "Strategic Quality Management," *Quality,* March 1990, pp. 17–22.
9. Doug Anderson, "The Role of Senior Management in Total Quality," *Global Perspectives on Total Quality,* Conference Board Report Number 958, New York: Conference Board, 1991, p. 17. The author is Director, Corporate Quality Services, 3M Company.
10. Lance Arrington, "Training and Commitment: Two Keys to Quality," *Chief Executive,* Sep. 1990, pp. 72–73.

11. Dianna Booher, "Link between Corporate Communication & Quality," *Executive Excellence,* June 1990, pp. 17–18.

12. Peter Drucker, *Management: Tasks, Responsibilities, Practices,* New York: Harper & Row, 1973, p. 481.

13. The author had the pleasure and learning experience of working with Larry Appley on the development of this program, which had great success in a number of companies.

14. Bruce Van-Lane, "Good as Gold," *PEM: Plant Engineering & Maintenance (Canada),* April 1991, pp. 26–28.

15. Erika Penzer, "Spreading the Gospel about Total Quality at a Westinghouse Symposium," *Incentive,* Feb. 1991, pp. 46–49.

16. Ralph H. Kilmann, Mary J. Saxton, and Roy Serpa, "Issues in Understanding and Changing Culture," *California Management Review,* Winter 1986, p. 89.

17. See "Jack Welsch Reinvents General Electric—Again," *The Economist (UK),* March 30, 1991. See also Joseph M. Juran, *Juran on Leadership for Quality: An Executive Handbook,* New York: The Free Press, 1989.

18. Edwards W. Deming, "Transformation of Today's Management," *Executive Excellence,* Dec. 1987, p. 8.

19. Armand V. Feigenbaum, "Seven Keys to Constant Quality," *Journal for Quality and Participation,* March 1989, pp. 20–23.

20. Philip B. Crosby, *Eternally Successful Organization,* New York: McGraw-Hill, 1988.

21. Aaron Sugarman, "Success through People: A New Era in the Way America Does Business," *Incentive,* May 1988, pp. 40–43.

22. Thomas C. Gibson, "The Total Quality Management Resource," *Quality Progress,* Nov. 1987, pp. 62–66.

23. Syed Shah and George Woelki, "Aerospace Industry Finds TQM Essential for TQS," *Quality,* March 1991, pp. 14–19.

24. U.S. General Accounting Office, *Quality Management Scoping Study,* Washington, D.C.: U.S. General Accounting Office, Dec. 1990, p. 64.

25. Tom Peters, "Total Quality Leadership: Let's Get It Right," *Journal for Quality and Participation,* March 1991, pp. 10–15.

26. Frederick Derrick, Harsha Desai, and William O'Brien, "Survey Shows Employees at Different Organizational Levels Define Quality Differently," *Industrial Engineering,* April 1989, pp. 22–26.

27. David A. Garvin, "Quality Problems, Policies, and Attitudes in the United States and Japan: An Exploratory Study," *Academy of Management Journal,* Dec. 1986, pp. 653–673.

28. David T. Kearns, "Leadership through Quality," *Academy of Management Executive,* Vol. 4 No. 2, 1990, p. 87.

29. The need to set activity output and productivity measures (productivity being the ratio of output to input) has long been recognized, but little progress was made until recently, when a method known as *activity analysis* began to emerge. It is still in the early stages of implementation in industry. The accounting profession, recognizing the need to focus not only on cost but also on quality and productivity, is now promoting a related method known as *activity-based accounting.* See H. Thomas Johnson, "A Blueprint for World-Class Management Accounting,"

Management Accounting, June 1988, pp. 23–30. A very insightful and useful treatment of the shortcomings of traditional accounting systems is contained in H. Thomas Johnson and Robert S. Kaplan, *Relevance Lost,* Boston: Harvard Business School Press, 1991. Also included are excellent prescriptions for the improvement of accounting systems for managerial decision making.

30. Joel K. Leidecker and Albert V. Bruno, "Identifying and Using Critical Success Factors," *Long Range Planning,* Vol. 17 No. 1, 1984, pp. 23–32.

31. John E. Newcomb, "Management by Policy Deployment," *Quality,* Jan. 1989, pp. 28–30.

32. Andrew D. Szilagyi, Jr. and Marc J. Wallace, Jr., *Organizational Behavior and Performance,* 5th ed., Glenview, Ill.: Scott, Foresman, 1990, p. 620.

33. Michael E. Porter, *Competitive Strategy,* New York: The Free Press, 1980, p. 40.

34. Peter Drucker, *Management: Tasks, Responsibilities, Practices,* New York: Harper & Row, 1973, p. 504.

3

INFORMATION and ANALYSIS

Since quality programs are dependent on good information systems, chief information officers have the opportunity to plan an integral and highly visible role in shaping the quality of the corporation.

Curt Reimann, Director
Malcolm Baldrige Award

Information is the critical enabler of total quality management (TQM). More and more successful companies agree that information technology and information systems serve as keys to their quality success. Conversely, this component of TQM is frequently the roadblock to improvement in many firms. In these firms better quality and productivity may not be the issue; rather, the real issue may be better quality of information. Dr. Curt W. Reimann, director of the Malcolm Baldrige National Quality Award, suggests that the critical constraint for many companies in applying for the award is the lack of a proper information system for tracking and improving areas in the remaining award categories.[1]

ORGANIZATIONAL IMPLICATIONS

John Sculley, former chairman of Apple Computer, concludes that information systems and technology can no longer be regarded as staff or service functions for management. Moreover, information systems will become the most important means for companies to create distinctive quality and unique service at the lowest possible cost.[2] At a 1988 symposium in Washington, D.C. for some 175 chief executive officers of major U.S. corporations, the main topic was quality improvement and the information systems to support that effort. Designing the product, the plant configuration, and even the organizational structure is less challenging than designing the information system, which is the central component of TQM that allows the process to function.[3] It may be that the rigor of the production process is not matched by that of the information system, and the cause may lie in the increased complexity and breadth of the latter. Information is critical to *all* functions, and *all* functions need to be integrated by information.

The natural progression of information systems (used interchangeably with management information systems) in the past has frequently resulted in "band-aids" or "islands of mechanization," as applications such as inventory control, production scheduling, and sales reporting were designed without much regard for integration among each other or among other functions and activities within the organization. In recent years, additional and more sophisticated applications have emerged, such as quality function deployment (QFD), Taguchi methods, statistical process control (SPC), and just-in-time (JIT). These are now considered basic to the TQM process.[4] The challenge remains the same: to integrate these techniques and principles into a structured approach that includes related decision-making requirements across the board.

In Chapter 6, the argument is made for designing processes for continuous improvement in quality and productivity. A natural accompaniment is the design of the information systems to facilitate decisions related to these improvements. Indeed, these modern processes are all but impossible without sophisticated information systems.

Historically, companies have automated the easy applications: payroll, financial accounting, production control, etc. Today, the concept of *re-engineering* is emerging. Rather than automating tasks and isolating them into discrete departments, companies are attempting to integrate the related activities of engineering, manufacturing, marketing, and support operations. Actions proceed in parallel, rather than sequential, order. Cycle time is reduced and products get to market faster with fewer defects. In short, the process is re-engineered, and computer power is applied to the new process in the form of information systems. The focus is changing from buying information technology in order to automate paperwork to a focus designed to improve the process.

Information Technology

Systems design may be a constraint, but information technology (IT) is not. The geometric acceleration of developments is well known and can only be described as dramatic and spectacular.[5] If industry is capable of absorbing the technology, a further increase in the sophistication and importance of information will occur. Capital and direct labor will continue to be sources of value added, but the proportion contributed by intellectual and information activity will increase. Indeed, information can be considered to be a substitute for other assets because it can increase the productivity of existing capital and reduce the requirement for additional expenditures. It should be exploited.[6]

In 1990, Federal Express spent more than $243 million on IT. CEO Fred Smith stated that IT is absolutely the key to the organization's operations and that the entire quality process depends on statistical quantification which, in turn, depends on IT. Information is generated for both employees and customers.

Decision Making

The ability to make decisions quickly has always been critical to management at all levels, and information is essential to the process. It has emerged as a crucial competitive weapon.[7] Yet middle managers, who are the real change agents, spend most of their time exchanging information with subordinates, peers, or the boss, leaving little time for customers or for innovation and change. In the jargon of information systems, they need a decision support system.

Information Systems in Japan

In what continue to be customary comparisons between the United States and Japan, it is useful to examine how IT and information systems are perceived in Japan. Japanese executives believe that customer satisfaction drives the development of new services and products and that IT can be a vital means to facilitate strategies and operations to this end.[8] In true Japanese fashion, this view is apparently promoted by the national government as well. To build a foundation for future technicians and managers, the Ministry of Education has implemented national education policies for the full-scale use of computers in education.[9] There is also a national policy on software. The Ministry of International Trade and Industry (MITI) has launched the Sigma Project, which calls for computerizing the software process and industrializing and computerizing software production.[10]

The Deming Prize is awarded each year to Japanese companies that demonstrate outstanding improvement in quality control. Yokogawa Hewlett-Packard (YHP), a joint venture of Hewlett-Packard and Yokogawa Electric Works, was awarded the prize for an information systems approach that yielded dramatic increases in profit, productivity, and market share.

STRATEGIC INFORMATION SYSTEMS

The integration of management information systems (MIS) with strategic planning has been suggested as a necessary prerequisite to strategy formulation and implementation. If we assume, as we must, that the basic requirement of a strategy is environmental positioning in order to meet customer requirements and if we further assume that the ultimate purpose of each function and process within an organization is to contribute to strategy, the role of information becomes clear.

As will be discussed in Chapter 9, the value chain is a useful concept for determining the structure and processes needed by organization in order to achieve a competitive advantage, keeping in mind that competitiveness is decided neither by the industry nor by the company, but rather by the customer.

Beginning with the customer, integration of processes and information can proceed as follows:

- Identify the market segment in which you want to compete.
- Use data collection and analysis to define the customer requirements in the chosen segment.
- Translate these requirements into major design parameters to develop, produce, deliver, and service the product that meets the customers' requirements. These are the primary functions and activities (processes) of the value chain.
- Complement the primary processes with support activities such as planning, finance and accounting, MIS, personnel, etc.
- Subdivide or "explode" the organization design parameters into the processes (functions, activities, etc.) that are necessary to achieve the quality differentiation.

▪ Design the information requirements necessary to manage each process and to integrate all processes horizontally.

The support activities are sometimes taken for granted and their linking potential is often overlooked. Moreover, their potential contribution to differentiation may not be realized. Marketing services, for example, when combined with the customer's expertise, can generate differentiated product and service opportunities. The customer will place high value on a supplier who delivers the right information quickly. Engineering services, usually perceived as a commodity product, can also differentiate a firm. In both cases the information systems support is cost effective.

At Honeywell, Inc., translating long-term strategy into tactics that enhance short-term operations has resulted in new approaches that have shortened cycle time, improved quality, and reduced costs. The approach involves spreading information, standardizing, and measuring performance.

Environmental Analysis

Strategy formulation requires an analysis of the different environments: general, industry, and competitive (see Chapter 4 for further discussion). One study found that small business owners spend over one-fourth of the day in external information search activities.[11] Competitive information is particularly valuable but is difficult to obtain.[12] In general, the minimum information needed about competitors can be related to how they stand on the key success factors for a market segment. These may differ by industry and segment but usually include the following:[13]

Market share	Growth rate
Product line breadth	Distribution effectiveness
Proprietary advantages	Price competitiveness
Age and location of facility	Capacity and productivity
Experience curve effects	Value added
R & D advantage and position	Cash throw-off

Porter has identified the information needed for positioning in an industry and in a chosen market segment, and his system is widely used. His categories are (1) intensity of rivalry, (2) bargaining power of buyers, (3) bargaining power of suppliers, (4) threat of substitution, and (5) threat of new entrants.[13] Each category includes a number of elements or subtopics that should be determined and tracked with some type of information system.

Central to all information relating to strategy formulation and implementation is the need to *define and measure* the concept of quality of product and service—as determined by the customer. This step is fundamental to positioning and subsequent follow up.

SHORTCOMINGS of ACCOUNTING SYSTEMS

Financial information is perhaps the most widespread indicator of performance, and for many firms is the only indicator. Critics of accounting systems claim that they do not really support the operations and strategy of the company, two dimensions in which quality plays a dominant role. Despite the widely held conclusion that we are in the information age, management accounting would probably be labeled inadequate by managers who seek to support company operations and strategy through quality improvement. This is increasingly evident in the "new" manufacturing environment which is characterized by the trends and implications listed in Table 3-1.

Accountant bashing is becoming increasingly popular in the management literature. The trend is symbolized by Harvard Business School Professor Robert Kaplan in his popular book *Relevance Lost*.[14] He concludes that today's accounting information provides little help in reducing costs and improving quality and productivity. Indeed, he suggests that this information might even be harmful. Peter Drucker, another critic, describes some of the shortcomings that are generally recognized:[15]

1. Cost accounting is based on a 1920s reality, when direct labor was 80 percent of manufacturing costs other than raw material. Today it is 8 to 12 percent and in some industries (e.g., IT) it is about 3 percent.

2. Non-direct labor costs, which can run up to 90 percent, are allocated in proportion to labor costs, an arbitrary and misleading system. Benefits of a process change are allocated in the same way.

Table 3-1 New Manufacturing Environment

Trend	Implication for quality
Focus on manufacturing strategy	Quality rapidly becoming the central competitive edge of strategy
Production of high-quality goods	Quality directly related to market share, growth, profits
Reduction of inventory levels by just-in-time (JIT) inventory	Reduction of costs associated with excess inventory
Tight schedules	Improves availability to customer, another competitive edge perceived as quality by the customer
Product mix and variety	Allows focus on strategy and market segmentation
Equipment automation	Provides justification for quality and productivity improvement
Shortened product life cycle	Provides opportunity to expedite market shifts and incorporate new technologies into the product, but imposes additional stress on the quality management program
Organizational changes	Responsibility for quality delegated to strategic business units and product managers
Information technology	Allows greater control of cost of quality, quality management, and cross-functional integration

3. The cost system ignores the costs of *non-producing,* whether this be downtime, stockouts, defects, or other costs of non-quality.

4. The system cannot measure, predict, or justify change or innovation in product or process. In other words, accounting measures direct or real costs and not benefits.

5. Accounting-generated information does not recognize linkages between functions, activities, or processes.

6. Manufacturing decisions cannot be made as *business* decisions based on the information provided by accounting. The system confines itself to measurable and objective decisions and does not address the intangibles.

Efforts are underway to make accounting a true management and business system. For example, Computer-Aided Manufacturing-International (CAM-I) is a cooperative effort by automation producers, multinational manufacturers, and accountants to develop a new cost accounting system. Even internal auditors are examining their new role in TQM.[16]

ORGANIZATIONAL LINKAGES

The importance of data linkages is illustrated by data on service calls, a primary source of measuring product field performance. These are an important source of information for design, engineering, manufacturing, sales, and service. One research study[17] reported that in some cases among air conditioner manufacturers, the aggregative data on failure rates was of little use because of organization barriers.

▪ The *service tracking report* at American Express monitors performance for all centers worldwide. For the credit card division, for example, performance is measured against 100 service measures, including how long it takes to process an application, authorize charges, bill card members, and answer customer billing inquiries. Each measure is based on customer expectations, the competition, the economy, and legislation. Application processing time has been reduced by 50 percent and the bottom line has been increased by $70 million.

This example illustrates the widespread need for organization linkages and cross-functional MIS and the need to track a process on a continuous basis. Figure 3-1 shows how each step in the life cycle of a product involves related processes as cross-functional lines.

Each step in the product life cycle involves a number of processes at these cross-functional lines in a continuous flow from design to preproduction planning to vendor management to incoming material to in-process control to finished goods to customer service. The steps along the flow should be accompanied by appropriate information.[18] Thus, the linkage concept may focus on internal customers (those who use products in a later step of the process) as well as external customers.

▪ Federal Express, the first service sector company to earn the Baldrige Award, integrates a variety of internal measurement systems into the core of its business. The objective is

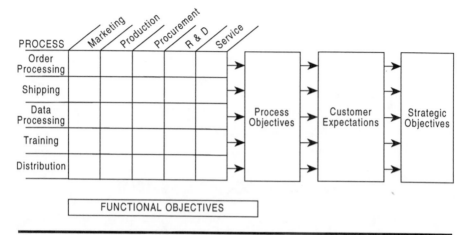

Figure 3-1 Cross-Functional Lines in the Life Cycle of a Product

"zero service defects." The system, SQI (service quality indicators) measures twelve critical points at which failure can occur in the service process and continually reinforces how employees are doing compared to their goals.

White-Collar Measures

The large number of white-collar and service personnel in the typical manufacturing firm was noted earlier. These activities not only comprise a large (perhaps major) share of total costs, but are essential to a systems approach to TQM. If TQM is to work, it must address the involvement of employees in developing measurement systems that will need to be in place and accessible to all levels.

The characteristics of white-collar work make it more difficult to measure than work in the manufacturing process. The American Productivity Center in Houston has developed a continuous performance improvement process for white-collar work called IMPACT.[19] *Measures development* is the fourth step of the six-step process. To quote:

> IMPACT provides a "family of measures" that allows each pilot group to track its progress from "Where are we?" to "Where do we want to go?" The family of measures provides the pilot group the tools it needs to measure progress, to give feedback, and to know when to take additional corrective actions. In addition, current measures are inventoried and used along with new measures. It is

most effective if both the customer and the supplier participate in this phase.

Sara Lee Hosiery launched a company-wide quality program that focuses on customer service and providing training designed to help information systems, networking, and other service departments understand how to apply total quality concepts to their work. The key is to focus on tools to enhance those processes that are based on customer needs and expectations.

ADVANCED PROCESSES/SYSTEMS

SPC...QFD...CAD...CIM...MRP—one gets the impression of "alphabet management." These and other basic applications represent the major systems of TQM. None stand alone and there are overlaps among them. Some advocates promote one or more as the "total system." Most if not all of these processes depend upon IT and a sophisticated information system design. Because systems design begins with the objective of the process, it is useful to list the objective of the major processes (Table 3-2).

At Motorola's Automotive and Industrial Electronics Group in Arcade, New York, over 1000 employees were trained in SPC. Operators then began doing their own inspections and plotting hourly control charts to control their own projects. Quality control inspectors were transferred out. Improvements included (1) achieving 10:1 goal of improvement, (2) significant increases in yields, and (3) reduction in scrap. The facility received the Q1 quality award as a supplier to Ford.

Donald Bell, general manager of Monsanto's Fibers Division, envisioned the "Plant of the 1990s." The scheme is a three-tiered approach encompassing human resources planning, total quality concepts, and computer-integrated manufacturing (CIM). Productivity gains of 40 to 50 percent have already been achieved. The program emphasizes the needs of internal customers—those who use products in a later step of the manufacturing process. Computer training has enabled greater acceptance of these concepts.[21]

INFORMATION and the CUSTOMER

According to examiners who visit companies that apply for the Baldrige Award, most companies lack the processes that ensure effi-

Table 3-2 Objectives of Major Processes in Systems Design

Process/system	Objective
Statistical quality control (SQC)	Build in the control limits of a process that spots and identifies causes of variations
Statistical process control (SPC)	Provide information on how productivity and quality can be *continuously improved* through problem identification (it has been estimated that U.S. firms invest 20 to 25 percent of their operating budgets in finding and fixing mistakes[20])
Just-in-time (JIT)	Reduce inventory cost, production time, and space requirements
Computer-integrated manufacturing (CIM)	Lower cost, shorten lead time, and improve quality based on information sharing by linking management and financial information systems, departmental computing, process management systems, and factory systems for controlling machinery and manufacturing processes
Quality function deployment (QFD)	Integrate the three dimensions of (1) company-wide quality, (2) focus on customer requirements, and (3) translation of quality perceptions into product characteristics and then into the manufacturing process

cient flow of information on customer demands and related information throughout the organization.[22] In other words, most companies do not devote the same attention to the customer that they do to the internal processes of shipping, inventory, JIT, manufacturing, etc. This is unfortunate because the operating processes cannot be managed according to the principles of TQM unless the loop is closed with customer feedback. Information systems should be extended beyond the plant into the marketplace. Some companies tend to define quality in terms of customer satisfaction or some other non-specific term and then relax after shipment is made, overlooking the competitive success that accompanies after-the-sale service, spare parts, or distribution.[23]

Why do information systems directly related to customer satisfaction frequently take a back seat to what otherwise might be acceptable or excellent information systems in support of quality and process control? The answer may be that it is difficult to specify information needs for an elusive system to measure customer requirements and satisfaction, which in themselves are difficult to define. Or it may be that the pressures of crisis management and internal information exchange leave little time for the customer.[24] Whatever the cause, it is a good idea to design a system that measures the pulse of the market and the customer base. It is estimated that failure to do so will cost twice as much as poor internal quality.[25]

> ■ The First National Bank of Chicago found that quality can be the difference between acquiring and keeping customers. Because competitive pricing varies by only a few pennies, the customer must be enlightened as to the benefits of strong quality. The bank measures customer satisfaction by how often inaccurate information is given. In 1982, the error rate was 1 in 4,000 transactions; in 1990 it dropped to 1 in 810,000 transactions.

Information Needs

After the objective of an information system is established, the next step is to determine the information needs. This is the most difficult step in designing a MIS for customer satisfaction. Everything else is detail and technique. Manager/user involvement is essential here.

If there is one fundamental principle of TQM, it is that *quality is what the buyer defines it as,* and not what the company defines it to be. Ford learned this lesson in the late 1970s, when the company definition of DQR (durability, quality, and reliability) was found to be presented in terms (engineering design tolerances and specifications) understand-able only within the company, rather than in terms that represented quality to the customer. Only after reassessing quality in terms under-standable to the customer was Ford able to adopt a policy called "Ford Total Quality Excellence" and achieve organization-wide commitment to continuous improvement and customer focus.

The first step then is to define quality as perceived by the customer by viewing it *externally* from the customer's perspective. By profiling how customers make purchase decisions, it is possible to determine which product attributes are most important and to determine how customers rate each attribute. As will be discussed in Chapter 8, this process forms the basis of *benchmarking.*[26]

Market research methods ranging from focus groups to shopper surveys are means for profiling customers and defining quality as perceived by customers. The information system can then be designed to provide the input for decisions regarding the operating plan, organizational implications, and follow-up control.

The INFORMATION SYSTEMS SPECIALISTS

It would be an understatement to say that the power of IT and computers has exploded and will continue to do so. Computer power is estimated to double every five years, while the cost continues to decrease. This expansion of IT and computer power has been accompanied by a growth in "knowledge workers" or "information workers"—people who control the quality of streams of information. These streams flow into a business from customers and the external environment, then flow through a business from product development to manufacturing and distribution, and flow out in the form of sales effort and service follow up.

It is unfortunate that the growth of the white-collar sector has outpaced its productivity. Still worse, while investment in IT tripled from 1978 to 1988, output per hour among 80 percent of the total workforce remained virtually stagnant. This represents a major opportunity to improve the productivity and quality of all workers by providing better decision-making information for process improvement throughout the company. Achieving this goal requires training in the techniques of systems design and use.

The Chief Information Officer

▪ General Dynamics was awarded the Premier 100 ranking as the top aerospace firm by the magazine *Computerworld*. To General Dynamics, TQM means near perfect products and an information system that meets the continuous process improvements. The chief information officer (CIO) of the company indicates that its real strength is having a clearly stated vision for information systems management.

Information systems is a growth industry, particularly for companies aspiring to TQM. Therefore, it is likely that quality information systems as described in this book will become more important and more voluminous than financial systems (the traditional and widespread source of operating data). This trend is reflected in the growing number of organizations with positions titled CIO.[27]

The CIO position is still evolving, but the ideal job description would have the individual responsible for developing IT/information systems planning and tying it into the strategic plan of the business (something that accounting does not generally do). Additionally, the CIO's function would focus on performance measures based on customer satisfaction and would then apply productivity tools to improve the related processes.[28] These functions would be in addition to the normal duties of quality assurance, cost-benefit analysis, software development, technology transfer, and technology forecasting. Providing quality output within the department to internal customers is a given because it sets a climate and provides a role model for others who deal with external customers. Service quality can only be built from the inside out, and how the information systems function delivers its services to internal customers can influence the way external customers are served.

SYSTEMS DESIGN

After reviewing hundreds of applications for the Baldrige Award, Curt Reimann, director of the award, concluded that the area of information and analysis represents a serious national problem.[29] Many firms have failed to design individual applications to fit an overall master plan. The result has been a "band-aid" solution with little integration between functions and activities.

A master plan should be centered around corporate goals and the critical success factors and cost-performance drivers related to these goals. In a manufacturing firm, data from engineering, production, and field service are used to improve product design and manufacturing techniques. If reducing cycle time in bringing a product to market is a critical success factor (as it is), a good deal of this information will flow sideways and across departmental lines, rather than upward and vertically as in the traditional model.

The individual manager/user has the job of designing his or her own system requirements and fitting these into the overall master plan. This is not easy. In discussions with dozens of system analysts, they almost always report that their number one difficulty in system design is the inability or unwillingness of the user to define information needs. This definition is not the job of the analyst—it is the job of the individual user. Before design can proceed, two critical steps must be taken: define *system objectives* and *information needs*.

Surprisingly, many users cannot define an objective. They will define it as "having the right part at the right place at the right time" or "preparing a field service report." Statements such as these are elusive, not quantifiable, and unsuitable for conversion to information needs. On the other hand, when objectives are stated in more specific terms (such as "reduce final inspection in the production process to the point of elimination" or "reduce throughput time to 6 days"), the designer has a benchmark from which to proceed.

The next step is to define *information needs,* another requirement that users have difficulty defining. The question is: "What information do I need to achieve the objective?" If performance measures are established, the determination of both objective and information needs will become more apparent. Successful companies benchmark their performance against world-class quality leaders. For example, Xerox measured its performance in about 240 key areas of product, service, and business performance. This process is discussed further in Chapter 8.

QUESTIONS for DISCUSSION

3-1 Describe how lack of information can be a roadblock to implementing one or more TQM actions.

3-2 How do traditional accounting systems provide inadequate information for control of processes in an industry with low labor content.

3-3 Choose two functions or activities (market research, R&D, design, production planning, procurement, human resources) and show how information can serve to integrate them across functional lines.

3-4 How does information technology affect organizational structure? Give an example of how information technology can facilitate TQM.

3-5 How would you go about designing an MIS for getting customer input for quality improvement?

3-6 How does market segmentation influence information needs?

ENDNOTES

1. Telephone interview with Curt W. Reimann.
2. John Sculley, "The Human Use of Information," *Journal for Quality and Participation,* Jan./Feb. 1990, pp. 10–13.
3. Elizabeth A. Haas, "Breakthrough Manufacturing," *Harvard Business Review,* March/April 1987, pp. 75–81. It is estimated here that companies adopting integrated strategies may succeed in increasing productivity by 10 or 15 percent annually. See also Julian W. Riehl, "Planning for Total Quality: The Information Technology Component," *Advanced Management Journal,* Autumn 1988, pp. 13–19.
4. Nael A. Aly, Venetta J. Maytubby, and Ahmad K. Elshennawy, "Total Quality Management: An Approach & A Case Study," *Computers and Industrial Engineering,* Issues 1–4, 1990, pp. 111–116.
5. For a description of what lies ahead, see Robb Wilmot, "Computer Integrated Management—The Next Competitive Breakthrough," *Long Range Planning,* Vol. 21 No. 6, 1988, pp. 65–70.
6. James Heskett, "Lessons in the Service Sector," *Harvard Business Review,* March/April 1987, pp. 118–126.
7. Kathleen M. Eisenhardt, "Speed and Strategic Choice: How Managers Accelerate Decision Making," *California Management Review,* Spring 1990, pp. 39–54.
8. Dennis Normile, "Japan Inc. Bows to the Customer," *CIO,* Aug. 1990, pp. 91–93. A major benefit of information systems is increasing the speed from product concept to marketing, an improvement that translates into customer satisfaction.
9. Takashi Yamagiwa, "Computer Use in the Japanese Educational System," *Business Japan,* March 1988, pp. 38–39.
10. Ryozo Hayashi, "National Policy on the Information Service Industry," *Business Japan,* March 1990, pp. 49–61.
11. J. Lynn Johnson and Ralph Kuehn, "The Small Business Owner/Manager's Search for External Information," *Journal of Small Business Management,* July 1987, pp. 53–60.
12. The pioneering book is Frank J. Aguilar, *Scanning the Business Environment,* New York: Macmillan, 1967.
13. For a comprehensive treatment of competitive information and its sources, see Michael E. Porter, *Competitive Advantage,* New York: The Free Press, 1985. Based on research data collected from more than 3000 strategic business units, The Strategic Planning Institute, through its PIMS program, has identified the following characteristics of the most profitable companies in an industry: (1) higher market share, (2) higher quality, (3) higher labor productivity, (4) higher capacity utilization, (5) newer plant and equipment, (6) lower investment intensity per sales dollar, and (7) lower direct cost per unit. See Robert D. Buzzell and Bradley T. Gale, *The PIMS Principles: Linking Strategy to Performance,* New York: The Free Press, 1987.
14. H. Thomas Johnson and Robert S. Kaplan, *Relevance Lost: The Rise and Fall of Management Accounting,* Boston: Harvard Business School Press, 1991.
15. Peter Drucker, "The Emerging Theory of Manufacturing," *Harvard Business Review,* May/June 1990, pp. 94–102. See also Robert S. Kaplan, "The Four Stage

Model of Cost Systems Design," *Management Accounting*, Feb. 1990, pp. 22–26; James M. Reeve, "TQM and Cost Management: New Definitions for Cost Accounting," *Survey of Business*, Summer 1989, pp. 26–30.

16. Fred J. Newton, "A 1990s Agenda for Auditors," *Internal Auditor*, Dec. 1990, pp. 33–39.

17. David A. Garvin, *Managing Quality*, New York: The Free Press, 1988, pp. 167–169. In this study, the best plants maintained sophisticated systems to track data and report it back to interested departments and functions.

18. See Raymond G. Ernst, "Why Automating Isn't Enough," *Journal of Business Strategy*, May/June 1989, pp. 38–42. The author argues that companies too often attempt to improve manufacturing by making large investments in automation without improving their business processes. He estimates that the savings of 10 to 20 percent that can be derived from automating can be increased to 70 percent when improvements are made to existing business processes as well. The processes can be achieved through a product-information flow.

19. Jackie P. Comola, "Designing a New Family of Measures," in *Total Quality Performance*, New York: The Conference Board, 1988, pp. 59–64. Mr. Comola is Vice President, White-Collar Productivity at the American Productivity Center. Measures are developed through a nominal group technique (NGT) in which teams accomplish the following steps: state the problem, list possible measures, round-robin collection of possibilities, edit nominations, vote and rank, discussion, and consensus. The six phases of IMPACT are (1) planning, (2) assessment, (3) direction setting, (4) development of measures, (5) service (re)design and implementation, and (6) results review and recycle.

20. Otis Port, "The Push for Quality," *Business Week*, June 8, 1987, pp. 130–135.

21. For an excellent description of how QFD is implemented, see John R. Hauser and Don Clausing, "The House of Quality," *Harvard Business Review*, May/June 1988, pp. 63–73. See also Chia-Hao Chang, "Quality Function Deployment (QFD): Processes in an Integrated Quality Information System," *Computers and Industrial Engineering*, Vol. 17 Issues 1–4, 1989, pp. 311–316.

22. Peter Burrows, "Commitment to Quality: Five Lessons You Can Learn from Award Entrants," *Electronic Business*, Oct. 15, 1990, pp. 56–58.

23. Morris A. Cohen and Hau L. Lee, "Out of Touch With Customer Needs? Spare Parts and After Sales Service," *Sloan Management Review*, Winter 1990, pp. 55–66.

24. Robert W. Wilmot, "Computer Integrated Management—The Next Competitive Breakthrough," *Long Range Planning*, Dec. 1988, pp. 65–70. This author has found that typical middle managers spend less than 10 percent of their time with customers and a tiny fraction sponsoring innovation and orchestrating change.

25. John Goodman and Cynthia J. Grimm, "A Quantified Case for Improving Quality Now," *Journal for Quality and Participation*, March 1990, pp. 50–55.

26. Bradley T. Gale and Robert D. Buzzel, "Market Perceived Quality: Key Strategic Concept," *Planning Review*, March/April 1989, pp. 6–15.

27. A recent survey by the Healthcare Financial Management Association found increasing use of a CIO in large hospitals and a growing emphasis on strategic planning. See John J. May and Ed H. Bowman, "Information Systems for the Value Management Era," *Healthcare Financial Management*, Dec. 1986, pp. 70–74. Insurance companies and banks represent other industries that are adopting this position.

28. Herbert Z. Halbrecht, "What's Good For the Boss...", *Computerworld,* Aug. 21, 1989, p. 78.
29. Curt W. Reimann, "Winning Strategies for Quality Improvement," *Business America,* March 25, 1991, pp. 8–11.

4

STRATEGIC
QUALITY PLANNING

*The basics of total quality management (TQM) can effectively
govern executive-level strategic management and goal-setting*

Executive
Academy of Management

Ford's slogan, "Quality Is Job 1," has caught on with increasing segments of the car-buying public. The company's North American Automobile Group is gaining market share among U.S. manufacturers and
has a higher net income than General Motors with only two-thirds the
amount of sales.[1] Things were not always this way. Between 1978 and
1982 market share slipped to 16.6 percent and sales fell by 49 percent,
with a cumulative loss in excess of $3 billion. Ford was losing $1000
on every car it sold. The company sought advice from W. Edwards
Deming. Reports John Betti, at that time a senior executive at Ford, "I
distinctly remember some of Dr. Deming's first visits. We wanted to talk
about quality, improvement tools, and which programs work. He
wanted to talk to us about management, cultural change, and senior
management's vision for the company. It took time for us to understand
the profound cultural transformation he was proposing."[2] The company's
subsequent turnaround is a classic example of the results that can be
obtained from a strategic change based on quality. The major changes
responsible for reversing the company's fortune were as follows:

▪ Emphasize quality and review new product planning and design.

▪ Keep investing in new products and processes.

▪ Make employee relations a source of competitive advantage.[3]

3M's approach to quality is so highly regarded that executives from leading U.S. companies travel to St. Paul to attend monthly briefings sponsored by 3M. In *Thriving on Chaos*,[4] Tom Peters described 3M as the only truly excellent company today. *Forbes* chose 3M as one of America's three most highly regarded companies. Their TQM implementation strategy includes:

▪ Defining 3M's quality vision

▪ Changing management perceptions through specialized training

▪ Empowering employees to focus on and satisfy customer expectations

▪ Sustaining the process through an ongoing culture change

One executive of the company explained it as follows: "How do you meet such a wide variety of expectations in a coherent way? I think you do it with a corporate philosophy on what constitutes a total quality process...a philosophy that you can apply across the company...to all your operations."[5]

These comments reflect the importance that successful companies place on the strategy issue. In the American Management Association's survey of over 3000 international managers, the key to competitive success was defined as the improvement of quality. There is little doubt that a strategy based on quality begins with strategic planning and is implemented through program and action planning.[6]

STRATEGY and the STRATEGIC PLANNING PROCESS

What is strategy and what is the strategic planning process? The answers to these questions are important because evidence suggests that those companies with strategies based on TQM have achieved stunning successes.[7]

Most of these successful companies will attribute their progress to a quality-based strategy that was developed through a formal structured approach to planning. The Commercial Nuclear Fuel Division of Westinghouse, another Baldrige winner, has discovered that the total quality concept must be viewed as a pervasive operating strategy for managing a business every day:

> Total Quality begins with a *strategic decision*—a decision that can only be made by top management—and that decision, simply put, is the decision to compete as a world-class company. Total Quality concentrates on quality performance—in every facet of the business—and the primary strategy to achieve and maintain competitive advantage. It requires taking a systematic look at an organization—looking at how each part interrelates to the whole process. In addition, it demands continuous improvement as a "way of life."[8]

Major contributors to the development of the strategic concept and to the planning process include Professors Andrews, Christensen, and others in the Policy group at the Harvard Business School.[9] A recent definition by this group is contained in their highly regarded text on the subject:[10]

> Corporate strategy is the pattern of decisions in a company that (1) determines, shapes, and reveals its objectives, purposes, or goals; (2) produces the principal policies and plans for achieving these goals; and (3) defines the business the company intends to be in, the kind of economic and human organization it intends to be, and the nature of the economic and non economic contribution it intends to make to its shareholders, employees, customers, and communities.

Michael Porter is perhaps the most highly regarded and certainly the most popular writer on the subject of strategy.[11] He describes the development of a competitive strategy as "a broad formula for how a business is going to compete, what its goals should be, and what policies will be needed to carry out those goals."

STRATEGIC QUALITY MANAGEMENT

This pattern of goals, policies, plans, and human organization is not something to be taken lightly. It is likely to be in place over a long period of time and therefore affects the organization in many different ways. The culture that guides members of the organization and other stakeholders, the position that it will occupy in an industry and market segments, and determining particular objectives and allocating resources to achieve them all follow from the decision processes determined by strategy. It is easy to see how pervasive a strategy based on quality can become. It provides the basis upon which plans are developed and communications achieved. A basic rule of strategic planning is that *structure follows strategy*. Although the process of formulation and implementation may require staff input, the ultimate decision is fundamental to the job of the chairman or CEO. It cannot be delegated.

The pervasive role that quality plays in strategic planning can best be understood by examining the components of a strategy:

- Mission
- Product/market scope
- Competitive edge (differentiation)
- Supporting policies
- Objectives
- Organizational culture

These components are developed through a process of strategy formulation, the outline of which is shown in Figure 4-1. Note that the process involves positioning yourself against forces in the environment in such a way that action plans can minimize your weaknesses and take advantage of your strengths relative to the competition. Quality is the means of differentiation for the satisfaction of customer needs. Research that includes over 300 U.S. companies indicates that firms with superior quality address quality offensively, as a distinct competitive advantage, while firms with inferior quality treat it defensively (e.g., eliminate defects, lower cost of product failure).[12]

Mission

The mission is the primary overall purpose of an organization and its expressed reason for existence. The simplest statement of mission might be to "meet the needs/values of constituents."

> ■ The mission of NCR is stated simply: "Create Value for Our Stakeholders." Stakeholders are identified as employees, shareholders, suppliers, communities, and customers.[13] The mission can be operationalized by statements of how it will be implemented for each stakeholder.

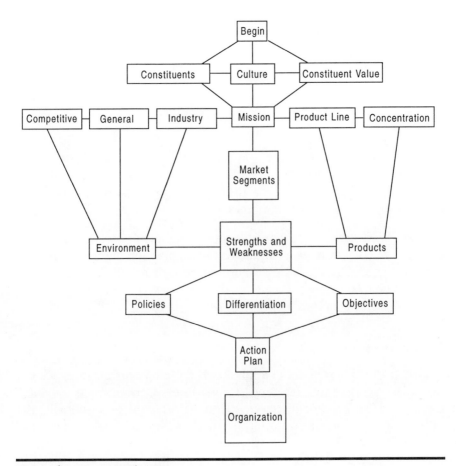

Figure 4-1 Strategic Planning

▪ At Goodyear, every employee carries a credit-card-sized mission statement: "Our mission is constant improvement in products and services to meet our customers' needs. This is the only means to business success for Goodyear and prosperity for its investors and employees."[14]

The mission statement includes the value that is being added and the direction the company intends to move. Because a mission can only be achieved by the people in an organization, it should have the commitment of the entire organization. Deming's first and what he considers his most important point of management obligation is to "create constancy of purpose for improvement of product and service with a plan to become competitive and stay in business."

This consistency must be achieved by a mission that can be operationalized and implemented. Consider the following examples:

■ All employees at Motorola consistently strive for a six sigma target.

■ 3M's mission focuses on innovation. To ensure consistency of purpose, the company established a requirement that 25 percent of each profit center's sales must come from products less than five years old.

■ Ford spent more than a year defining its mission. The real test of consistency and commitment came when the company withheld releasing a new Thunderbird, a "sure bet" for car of the year, because the car's quality was not yet suitable for a production model.

Environment

The major determinant of a mission is the environment in which the firm plans to operate: the general environment, the industry environment, and the competitive environment. Strategy is essentially the process of positioning oneself in that environment as trends and changes unfold. Thus, it is necessary to identify trends in the environment and how they affect the strategy of the firm. Figure 4-2 illustrates how a major U.S. manufacturer of computer equipment and software documented the major changes in that industry. The impact on strategy, as these issues relate to quality, is illustrated in Figure 4-3.

Product/Market Scope

This answers the questions: What am I selling and to whom am I selling it? The answers are more complex than they appear. What is Domino's Pizza selling: dough and tomato sauce or reliable delivery? What is a physician selling: surgery and diagnosis or patient involvement? Wal-Mart and Bloomingdale's are both in the retail business, but are their products simply what is on the shelves and racks in their stores? A company does not simply sell shoes or soap or banking services. It sells value to a particular segment of the market. The

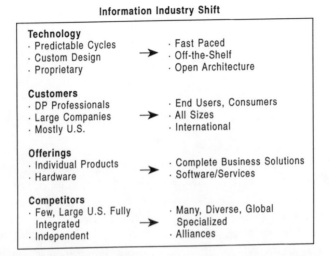

Information Industry Shift

Technology
· Predictable Cycles → · Fast Paced
· Custom Design → · Off-the-Shelf
· Proprietary → · Open Architecture

Customers
· DP Professionals → · End Users, Consumers
· Large Companies → · All Sizes
· Mostly U.S. → · International

Offerings
· Individual Products → · Complete Business Solutions
· Hardware → · Software/Services

Competitors
· Few, Large U.S. Fully → · Many, Diverse, Global
  Integrated   Specialized
· Independent → · Alliances

Figure 4-2   Defining the Environment

answers to these questions should be clear, as well as the role of quality in customer value.

What is value? It is, of course, what the customer—not the company—says it is. Timex sells watches, but does Rolex sell jewelry and prestige? Canada Dry sells sparkling water, but does Perrier sell snob

Re-Balancing for the '90s

Technology-Driven		Market-Driven
The Product	Focus	The Customer
Creating Demand	Approach	Providing Solutions
Price and Product Function	Marketing Strategy	Customer's Hearts and Minds
Product Volumes Revenue Profit	Measure of Performance	Customer Satisfaction Market Share Financial Returns

Figure 4-3   Impact of Changes in the Environment on Strategy

appeal? Thom-McCann sells loafers, but what does Gucci sell? This does not mean that Timex, Canada Dry, and Thom-McCann do not sell on the basis of quality. Indeed, they do. However, quality is defined differently for a different segment of the market. Each company must define its market segment and customer value in that segment. Ford's product mix includes the Lincoln Town Car and the Escort, but each is targeted at a different market segment, and quality (value) is different for each segment.

Every purchase decision is a function of price and quality. Price is generally known, but quality is in the mind of the individual customer. General Electric is aware of this and has broadened its perspective from "product quality" to "total customer satisfaction." The "product" is now defined by the customer.[15] It only remains to define customer satisfaction, perception, or expectation.

To repeat, in today's heightened competitive environment, a product or service is not simply sold to anyone who will buy it. To be effective, value must be sold to a particular market or customer segment. Strategic planning involves the determination of these strategy components, and quality plays a major role in this process.

Differentiation

Differentiation, frequently called the competitive edge, answers the question: Why should I buy from you? Michael Porter, in his landmark book *Competitive Strategy,* identified two generic competitive strategies: (1) overall cost leadership and (2) differentiation.[16] Cost leadership in turn can be broad in market scope (e.g., Ivory Soap, Emerson Electric, Black & Decker) or market segment focused (e.g., La Quinta Motels, Porter Paint). The second strategy involves differentiating the product or service by creating something that is perceived by the buyer as unique. Differentiation can also be broad in scope (American Airlines in on-time service, Caterpillar for spare parts support) or focused (e.g., Godiva chocolates, Mercedes automobiles). Thus, there are four generic strategies, but each depends on something different—something unique or distinguishing. Even an effective cost leadership strategy must start with a good product.

Selecting a strategy and recognizing quality as the competitive dimension is important for strategic purposes. Product and service quality has become widely recognized as a major force in the competitive marketplace and in international trade.[17]

Research indicates that eight out of ten customers consider quality to be equal to or more important than price in their purchase decisions.[18] This is a doubling of buyer emphasis in ten years and the trend is expected to continue. The message here is that whether a cost leadership or differentiation strategy is chosen, quality must be a competitive consideration in either case.

Differentiation can command a premium price or allow increased sales at a given price. Moreover, differentiation is one of two types of competitive advantage, the other being price. Price, however, should not be the sole basis of differentiation unless the product is perceived to be a commodity. Even if the product is a commodity or near-commodity, it can still be differentiated by such service characteristics as availability or cycle time.

The several sources of differentiation are not well understood. Many managers perceive their uniqueness in terms of the physical product or in their marketing practices rather than in terms of value to the customer. They may waste money because their uniqueness does not provide real value to the buyer. Why spend money on extra tellers or checkout lines to reduce waiting time to one minute if the customers are willing to wait two minutes? On the other hand, buyers frequently have difficulty estimating value and how a particular firm can provide it. This incomplete knowledge can become an opportunity if the firm can adopt a new form of differentiation and educate the buyers to value it.

DEFINITION of QUALITY

The concept and vocabulary of quality are elusive. Different people interpret quality differently. Few can define quality in measurable terms that can be operationalized. When asked what differentiates their product or service, the banker will answer "service," the health care worker will answer "quality health care," the hotel or restaurant employee will answer "customer satisfaction," and the manufacturer will simply answer "quality product." When pressed to provide a specific definition and measurement, few can do so.[19] There is an old maxim in management which says, "If you can't measure it, you can't manage it," and so it is with quality. If the strategic management system and the competitive advantage are to be based on quality, every member of the organization should be clear about its concept, definition, and measurement as it applies to his or her job. As will be discussed, it may be entirely appropriate for quality to be defined or perceived differently in the same company depending on the particular phase of the product life cycle.

Harvard professor David Garvin, in his book *Managing Quality*,[20] summarized five principal approaches to defining quality: transcendent, product based, user based, manufacturing based, and value based.

People from around the world travel to view the Mona Lisa or Michaelangelo's David, and most would agree that these works of art represent quality. But can they define it? Those who hold the *transcendental* view would say, "I can't define it, but I know it when I see it." Advertisers are fond of promoting products in these terms. "Where shopping is a pleasure" (supermarket), "We love to fly and it shows" (airline), "The great American beauty...It's elegant" (automobile), and "It means beautiful eyes" (cosmetics) are examples. Television and print media are awash with such undefinable claims and therein lies the problem: quality is difficult to define or to operationalize. It thus becomes elusive when using the approach as a basis for competitive advantage. Moreover, the functions of design, production, and service may find it difficult to use the definition as a basis for quality management.

Product-based definitions are different. Quality is viewed as a quantifiable or measurable characteristic or attribute. For example, durability or reliability can be measured (e.g., mean time between failure, fit and finish), and the engineer can design to that benchmark. Quality is determined objectively. Although this approach has many benefits, it has limitations as well. Where quality is based on individual taste or preference, the benchmark for measurement may be misleading.

User-based definitions are based on the idea that quality is an individual matter, and products that best satisfy their preferences (i.e., perceived quality) are those with the highest quality. This is a rational approach but leads to two problems. First, consumer preferences vary widely, and it is difficult to aggregate these preferences into products with wide appeal. This leads to the choice between a niche strategy (see later) or a market aggregation approach which tries to identify those product attributes that meet the needs of the largest number of consumers.

Another problem concerns the answer to the question: "Are quality and customer satisfaction the same?" The answer is probably not. One may admit that a Lincoln Continental has many quality attributes, but satisfaction may be better achieved with an Escort. One has only to recall the box office success of recent motion pictures that suffer from poor quality but are evidently preferred by the majority of moviegoers.

Manufacturing-based definitions are concerned primarily with engineering and manufacturing practices and use the universal definition of "conformance to requirements." Requirements, or specifications, are established by design, and any deviation implies a reduction in quality. The concept applies to services as well as products. Excellence in quality is not necessarily in the eye of the beholder but rather in the standards set by the organization. Thus, both Cadillac and Cavalier possess quality, as do Zayre and Bloomingdale's, as long as the product or service "conforms to requirements."

This approach has a serious weakness. The consumer's perception of quality is equated with conformance and hence is internally focused. Emphasis on reliability in design and manufacturing tends to address cost reduction as the objective, and cost reduction is perceived in a limited way—invest in design and manufacturing improvement until these incremental costs equal the costs of non-quality such as rework and scrap. This approach violates Crosby's concept of "quality is free" and is examined further in Chapter 11.

Value-based quality is defined in terms of costs and prices as well as a number of other attributes.[21] Thus, the consumer's purchase decision is based on quality (however it is defined) at an acceptable price. This approach is reflected in the popular *Consumer Reports* magazine which ranks products and services based on two criteria: quality and value. The highest quality product is not usually the best value. That designation is assigned to the "best-buy" product or service.

Which Approach(es)?

Which definition or concept of quality should be adopted? If each function or department in the company is allowed to pursue its own concept, potential conflicts may occur:

Function	Quality concerns
Marketing	Performance, features, service, focus on customer concerns User-based concerns that raise costs
Engineering	Specifications Product-based concerns
Manufacturing	Conformance to specifications Cost reduction

Adopting a single approach could lead to cost increases as well as customer dissatisfaction. Each function has a role to play, but it cannot be played in isolation. A blend is needed to coordinate meeting each of the concerns listed.

Market Segmentation (Niche) Quality

Quality means different things to different people. In terms of strategic quality management, this means that the firm must define that segment of the industry, that generic strategy, and that particular customer group which it intends to pursue. This can be called a segmented quality strategy. The big three automobile manufacturers have wide product lines, each of which is marketed to a different part of the market and each with differing quality attributes.

Recent efforts to codify the concepts of quality and provide baselines for measurement have yielded the characteristics listed in Table 4-1. None of these dimensions stands alone. Differentiation may depend on one or more or a combination, but the point is that when differentiating based on quality, quality must defined in terms that meet

Table 4-1 Measurement of Quality

Category	Example
Performance	On-time departure of aircraft Acceleration speed of automobile
Features	Remote control for stereo Double coupons at the supermarket
Reliability	Absence of repair during warranty 30-minute pizza delivery
Conformance	Supplier conforms to specifications Cost of performance failures
Durability	Maytag's 10-year transmission warranty Mean time between failures
Serviceability	Consumer "hot line" for repair information Time to answer the telephone for reservation or complaint
Aesthetics	Restaurant ambiance Perfume fragrance
Perceived quality	Japanese vs. American automobiles Doctor A is better than Doctor B

Table 4-2 Factors in
Decisions to Purchase

Factor	Mean
Performance	9.37
Lasts a long time	9.03
Easy to repair	8.80
Service	8.62
Warranty	8.13
Price	8.11
Ease of use	8.09
Appearance	7.54
Brand name	6.09

customer expectations, even if this is only what the customer perceives as quality.

A survey of purchasers of consumer products by the American Society for Quality Control summarized the factors influencing decisions to purchase (on a scale of 1 to 10) (Table 4-2).

Objectives

Management statesman Peter Drucker has said, "a company has but one objective: to create a customer." Following this statement, he proceeded to popularize the concept of management by objectives (MBO) and identified eight key areas for which objectives must be set:[22] (1) marketing, (2) innovation, (3) human organization, (4) financial resources, (5) physical resources, (6) productivity, (7) social responsibility, and (8) profit requirements. These areas have been widely adopted by industry.

Within these eight broad areas, a company can set more specific objectives to identify the ends it hopes to achieve by implementing a strategy. Marketing becomes market share, innovation becomes new products, financial resources becomes capital structure, productivity becomes output per employee, profitability becomes return on investment or earnings per share, and so on.

Here, the question of quality becomes blurred. Is it a mission or an objective? It hardly matters if it is woven into the fabric of company strategy. If quality is chosen as central to a mission, other objectives begin to fall into place. For example, cycle time reduction, cost reduction, competitive standing, and return to shareholders can be related to the central mission.

■ Digital Equipment Corporation launched a TQM program in order to tie together various efforts scattered throughout the company. The goal is to have a consistent company vision and language. Included is a six sigma program motivated by a desire to improve competitive position.[23]

Many quality improvement programs were started in the 1980s and 1990s in reaction to the increasing importance of quality and the need to compete for market share. Many companies failed, often because they had no action plan for implementing a strategy that was based on objectives, a prerequisite for follow-on operational planning.[24]

Supporting Policies

Policies are guidelines for action and decision making that facilitate the attainment of objectives. Taken together, a company's policies delineate its strategy fairly well. Tell me your policies and I can tell you your strategy.

The role of policies as a critical element of strategy is displayed in Figure 4-4, which can be called the *policy wheel*. In the center are the mission (the purpose of the organization), the differentiation (how to compete in the market), and the key objectives of the business. The spokes of the wheel represent the functions of the business. Each function requires supporting policies (functional strategies) to achieve the *hub*. If the firm's strategy calls for competing on quality, then this becomes the impetus for policy determination. Each functional policy supports this central strategy and the objectives that are determined during the planning process.

A firm's policy choices are essential as drivers of differentiation. They determine what activities to perform and how to perform them. Grey Poupon's advertising policy for its premium mustard sets the product apart. Bic Pen's manufacturing policy of low-cost automation supports its low price. Avon's door-to-door distribution policy sets it apart. McDonald's policy of strict franchisee training and control allows it to retain its quality image. An airline's policy of "answering the phone on the third ring" reinforces a competitive edge of service.

Figure 4-4 Policy Wheel

Testing for Consistency of Policies

Assuming that the company has decided to make quality the central focus of its strategy, objectives are then set for profitability, growth, market share, innovation, productivity, etc. The test for consistency of supporting policies for a hypothetical firm is provided in Table 4-3. Of course, each policy is related to the hub and radiates from it. Like a wheel, the spokes must be connected.

CONTROL

The propensity of the American manager to focus on short-term financial goals is well known. In its simplest and most prevalent form, the control system consists of setting financial standards (the budget), getting historical feedback on performance (the variance report), and trying to meet targets after deviations have occurred.

Much has been written about the shortcomings of this approach. The major problem is the lack of focus on productivity (absolute, not financial measures), quality, and other strategic issues.

A system to control quality objectives, as distinct from quality on the shop floor, requires measures and standards designed for that purpose. Indeed, Juran suggests that the traditional control process may be put

Table 4-3 Consistency of Supporting Policies

Function	Illustration of policy
Target market	Map the industry and seek out those segments where we have the advantage
Product line	Product line breadth is confined to those products where our value chain is appropriate for focus segment
Marketing	Market research to be directed toward defining customer expectations
Sales	Sales force hired and trained to promote our competitive edge
Distribution	Select distributors that complement our quality edge
Manufacturing	Invest in automation for improvement of quality and productivity
Supplier	Select suppliers that have applied for Baldrige Award Make life contracts
Human resources	Require skill and experience level for new hires Partnership relations with union
Research and development	Percentage of budget devoted to quality improvement Products designed for ease of repair
Finance	Service procedures in billing activity Financial arrangements with suppliers

on hold while increasing the emphasis on quality planning and improvement.[25] Thus, planning and control of quality come together in an integrated system. The focus is on quality improvement set out in the planning process. The difference between traditional dollar accounting budgeting and the control of quality objectives is the participation of those who set standards and targets. Each function, department, or individual sets targets and provides real-time feedback as operations unfold.

SERVICE QUALITY

The differences between service and product quality were discussed in Chapter 1. This topic will be examined further in Chapter 7 (Customer Focus and Satisfaction). It is both more difficult and yet simpler to plan and control service quality than it is to plan and control product quality. It is more difficult because measurement is elusive and production is frequently one-on-one. Like product quality, service quality should live up to expectation, but this can be a pitfall if too much service is promised.

Service quality may be more easily planned, provided objectives are defined and people committed. In any case, the payoff can be years away, and no service can overcome other weaknesses in a business.[26] The system for quality service also requires new approaches, such as restructuring incentives. In any case, a good beginning approach can be based on the Baldrige Award criteria, which are the same for both product and service. Process control in service industries is discussed further in Chapter 6.

SUMMARY

Quality has taken center stage as the main issue in both national and corporate competitive strategies. Those organizations that adopt quality as a differentiation and a way of organizational life will, over the longer term, pull ahead of competition. Achieving this goal is not easy. It is more than just issuing pronouncements and engaging in company promotion.

When an organization chooses to make quality a major competitive edge, it becomes the central issue in strategic planning—from mission to supporting policies. An essential idea is that the product is customer value rather than a physical product or service. Another concept that is basic to the process is the need to develop an organizational culture based on quality. Finally, no strategy or plan can be effective unless it is carefully implemented.

QUESTIONS for DISCUSSION

4-1 Assume that an airline, a hotel, and a hospital have chosen quality for differentiation. Identify two or more measures of quality for a firm in each of these industries.

4-2 Illustrate a definition of
   ▪ *Transcendental* quality
   ▪ Product-based quality
   ▪ User-based quality
   ▪ Value-based quality

4-3 Choose an industry and a product or service within that industry and show how quality may differ for different segments or customer groups within that industry.

4-4 Is the objective of cost reduction in conflict with quality improvement? If so, illustrate how.

4-5 How can quality be reflected in the following:

- ▪ Distribution policy
- ▪ Human resources
- ▪ Sales
- ▪ Suppliers

4-6 Illustrate how trends in an industry can change a company's strategy.

ENDNOTES

1. United States General Accounting Office, *Quality Management: Scoping Study,* Washington, D.C.: U.S. General Accounting Office, Dec. 1990, p. 67.
2. United States General Accounting Office, *Quality Management: Scoping Study,* Washington, D.C.: U.S. General Accounting Office, Dec. 1990, p. 15.
3. The details of Ford's transformation are contained in HBS Case 390-083, available from HBR Publications, Harvard Business School, Boston, MA 02163. See also Richard T. Pascale, *Managing on the Edge,* New York: Simon & Schuster, 1990.
4. Tom Peters, *Thriving on Chaos: Handbook for a Management Revolution,* New York: Knopf, 1987.
5. Remarks of A. F. Jacobson at the Conference Board Quality Conference in Dallas, April 2, 1990.
6. Eric Rolf Greenberg, "Customer Service: The Key to Competitiveness," *Management Review,* Dec. 1990, pp. 29–31.
7. J. M. Juran, "Made in USA—A Quality Resurgence," *Journal for Quality and Participation,* March 1991, pp. 6–8.
8. "Performance Leadership through Total Quality," a presentation made to the Conference Board Quality Conference, April 2, 1990. Two other Westinghouse divisions were runners-up for the Baldrige Award in 1989 and 1990.
9. Kenneth R. Andrews, *The Concept of Corporate Strategy,* New York: Dow Jones-Irwin, 1971.
10. Joseph L. Bower, Christopher A. Bartlett, C. Roland Christensen, Andrall E. Pearson, and Kenneth R. Andrews, *Business Policy: Text and Cases,* 7th ed., Homewood, Ill.: Irwin, 1991, p. 9.

11. Michael Porter, *Competitive Strategy: Techniques for Analyzing Industries and Competitors,* New York: The Free Press, 1980. See also his *Competitive Advantage: Creating and Sustaining Superior Performance,* New York: The Free Press, 1985, and *The Competitive Advantage of Nations,* New York: The Free Press, 1990.

12. Joel Ross and David Georgoff, "A Survey of Quality Issues in Manufacturing: The State of the Industry," *Industrial Management,* Jan./Feb. 1991.

13. Company brochure entitled "NCR Mission."

14. U.S. General Accounting Office, *Quality Management: Scoping Study,* Washington, D.C.: U.S. General Accounting Office, Dec. 1990, p. 23. T. Boone Pickens, the quintessential LBO raider, was not very charitable to Goodyear. In a speech to the Strategic Planning Institute in Boston on October 23, 1989, he used Chairman Robert Mercer as an example of corporate America in the early 1980s: "bloated, uncompetitive, bureaucratic and barely accountable to anyone...what I call the BUBBA syndrome."

15. Elyse Allan, "Measuring Quality Costs: A Shifting Perspective," a presentation made to the Conference Board Quality Conference, April 2, 1990. *Global Perspectives on Total Quality,* Report Number 958, New York: The Conference Board, 1990, p. 35.

16. Michael Porter, *Competitive Strategy: Techniques for Analyzing Industries and Competitors,* New York: The Free Press, 1980, pp. 35–37.

17. J. M. Juran, "Strategies for World-Class Quality," *Quality Progress,* March 1991, pp. 81–85.

18. Armand V. Feigenbaum, "How to Implement Total Quality," *Executive Excellence,* Nov. 1989, pp. 15–16.

19. Y. K. Shetty and Joel Ross, "Quality and Its Management in Service Businesses," *Industrial Management,* Nov./Dec. 1985, pp. 7–12; Joel Ross and Y. K. Shetty, "Making Quality a Fundamental Part of Strategy," *Long Range Planning (UK),* Feb. 1985, pp. 53–58.

20. David A. Garvin, *Managing Quality,* New York: The Free Press, 1988, pp. 40–46.

21. In a survey of consumers' purchasing decisions conducted by the Gallup Organization, consumers were asked to rank (on a scale of 1 to 10) the importance of selected factors in the decision to purchase; 42 percent ranked price as 10. Other factors ranked as 10 were performance (72 percent), lasts a long time (58 percent), easily repaired (52 percent), service (50 percent), warranty (48 percent), ease of use (37 percent), appearance (28 percent), and brand name (15 percent). See *'88 Gallup Survey of Consumers' Perceptions Concerning the Quality of American Products and Services,* Milwaukee: American Society for Quality Control, 1988, p. 9.

22. Peter F. Drucker, *Management: Tasks, Responsibilities, Practices,* New York: Harper & Row, 1973, p. 100.

23. Rick Whiting, "Digital Strives for a Consistent Vision of Quality," *Electronic Business,* Nov. 26, 1990, pp. 55–56.

24. A. Blanton Godfrey, "Strategic Quality Management," *Quality,* March 1990, pp. 17–22.

25. J. M. Juran, "Universal Approach to Managing for Quality," *Executive Excellence,* May 1989, pp. 15–17. See also Bradley Gale and Donald J. Swmre, "Business Strategies that Create Wealth," *Planning Review,* March/April 1988, pp. 6–13.

Traditional strategic planning based on financial measures is being called into question because they do not look beyond more important measures such as quality.

26. David Eva, "The Myth of Customer Service," *Canadian Business,* March 1991, pp. 34–39.

5

HUMAN RESOURCE DEVELOPMENT and MANAGEMENT

At the heart of Total Quality Management (TQM) is the concept of intrinsic motivation. Empowerment—involvement in decision making—is commonly viewed as essential for assuring sustained results.

Healthcare Forum

Kaizen is a Japanese concept that means *continuous improvement.* Despite the perception of many U.S. managers that kaizen is not appropriate for American firms, there is abundant evidence that the concept is entirely in keeping with American values and norms. The approach offers a substantial potential for improvement if accompanied by an appropriate human resources effort. Indeed, it is becoming a maxim of good management that *human factors* are the most important dimension in quality and productivity improvement. People really do make quality happen.

Chief executive officers of some of America's most quality-conscious companies are quick to point out that the best way to achieve organization success is by involving and empowering employees at all levels. Some even say that employee empowerment is a revolution that will turn top-down companies into democratic workplaces.

> The whole employee involvement process springs from asking all your workers the simple question, "What do you think?"
>
> Donald Peterson
> Former Chairman of Ford

> To get every worker to have a new idea every day is the route to winning in the '90s.
>
> John Welch, Chairman
> General Electric

> The teams at Goodyear are now telling the boss how to run things. And I must say, I'm not doing a half-bad job because of it.
>
> Stanley Gault
> Chairman

Recall W. Edwards Deming's fourteen points discussed in Chapter 1. The basis of his philosophy is contained in the following principles: (1) institute training on the job, (2) break down barriers between departments to build teamwork, (3) drive fear out in the workplace, (4) eliminate quotas on the shop floor, (5) create conditions that allow employees to have pride in their workmanship and abolish annual reviews and merit ratings, and (6) institute a program of education and self-improvement.

Total quality management (TQM) has far-reaching implications for the management of human resources. It emphasizes self-control, autonomy, and creativity among employees and calls for greater active cooperation rather than just compliance.

INVOLVEMENT:
A CENTRAL IDEA of HUMAN RESOURCE UTILIZATION

■ Back in 1987 the Ames Rubber Corporation decided to adopt a TQM strategy as a major change for implementing their determination to become more competitive. The executive committee identified its best and brightest managers and asked them to reorganize around functional processes. By 1992, every employee was assigned to an *involvement* group or team.

The human resource professional magazine *HR Focus* asked over 1000 readers to rate the key issues they faced in 1993. Employee involvement was rated as one of the top three concerns by 46 percent of the respondents. Customer service followed with 39 percent and TQM with 34 percent.[1]

At the heart of TQM is the concept of intrinsic motivation-involvement in decision making. Employee involvement is a process for *empowering* members of an organization to make decisions and to solve problems appropriate to their levels in the organization. The logic is that the people closest to a problem or opportunity are in the best position to make decisions for improvement if they have ownership of the improvement process. Empowerment is equally effective in service industries, where most frequently the customer's perception of quality stands or falls based on the action of the employee in a one-on-one relationship with the customer.

At Federal Express the driver represents the company. He or she *is* the company and must deal directly with customer problems. Quality in an airline is represented not by CEOs and pilots, but by counter personnel and flight attendants.

■ One of the more successful efforts to *empower* employees was the Astronautics Groups at Martin Marietta's Denver, Colorado operation (MMAG). The group instituted a TQM process. To build employee support, the group dropped its pyramid hierarchy of management in favor of a flatter structure and a more participative management approach. High-performance work teams were organized to *empower* people closest to the work to make decisions about how the work is performed. Aside from the substantial production area savings, less tangible benefits included improved morale.

Quality improvement can result from a reduction in cost or cycle time, an increase in throughput, or a decrease in variation within the process. In the past, the focus in achieving such improvement was frequently the *system*—traditional techniques and methods of quality control. Such a focus may overlook the fact that operation of the system depends on people, and no system will work with disinterested or poorly trained employees. The solution is simple: coordinate the system and the people.

Contrast two production management styles in manufacturing industries. The "buffered" approach is characterized by large stocks of inventory and narrowly specialized workers. "Lean" systems, utilizing just-in-time (JIT) techniques, operate with small inventory stocks, multiskilled workers, and a team approach to work organization. Lean plants are more productive because they do not have valuable resources tied up in idle inventory. Plants are smaller and more efficient, with increased communication among departments, and workers tend to have a view of the organization as a whole.

Two examples of the lean approach involving worker participation are General Motors' New United Motor Manufacturing (NUMMI) plant (a joint venture with Toyota) and Dynatech's automotive test division. In both companies, *internalization* of the JIT philosophy and worker participation have increased worker pride and involvement on the shop floor. At GM, productivity levels are 40 percent higher than typical GM plants, and the plant has the highest quality levels GM has ever known. At Dynatech, cycle time was reduced by as much as 90 percent and setup by 67 to 100 percent.

ORGANIZING for INVOLVEMENT

Human resource professionals generally agree that a major shortcoming of human resource programs is the lack of recognizing how to match employee talent with organizational effectiveness. A strategy of empowerment must be operationalized through some organizational vehicle. A suggestion system is certainly not the total answer, despite the fact that many companies consider it to be an employee involvement program, and in many cases it is the only program.

Properly organized and administered small groups and teams are an effective motivational device for improvement of productivity and quality. They can reduce the overlap and lack of communication in a functionally based classical structure characterized by chain of command, territorial battles, and parochial outlooks. The danger always exists that functional specialists may pursue their own interests at the expense of the overall company mission or strategy. Team membership, particularly in a cross-functional team, reduces many of these barriers and encourages an integrative systems approach to achievement of common objectives—those that are common to both the company and the team or group. Consider the following success stories:

   ■ Globe Metallurgical, the first small company to win the Baldrige Award, had a 380 percent increase in productivity which was attributed to self-managed work teams.

■ Ford increased productivity by 28 percent by using the "partnering" concept that required a new corporate culture of participative management.

■ At Decision Data Computer Corporation, middle management was trained to support "Pride Teams."

■ Martin Marietta Electronic and Missiles Group achieved success with performance measurement teams (PMTs).

Quality circles are perhaps the most widespread form of employee involvement teams. They are defined as a small group of employees doing similar or related work who meet regularly to identify, analyze, and solve product quality and production problems and to improve general operations. The concept has had some success in white-collar operations, but the major impact has been among "direct labor" employees in manufacturing, where concerns focus primarily on quality, cost, specifications, productivity, and schedules. Few cross-functional problems are considered because problem solving is generally confined to similar work areas.

Quality circles have not met the expectations that were set for them. As many as 50 percent of Fortune 500 companies have disbanded their circles. The major reason has been a general lack of commitment to the concept of participation and involvement and the lack of interest by management. Many middle managers perceived quality circles as a threat to their power and authority.

Task teams are a modification of the quality circle concept. The major differences are that the task teams can exist at any level and the goal is given to the team, whereas quality circles are generally free to choose the problems they will address.

Self-managing work teams are also an extension of the quality circle concept but differ in one major respect: members are empowered to exercise control over their jobs and optimize the effectiveness of the total process rather than the individual steps within it. Team members perform all the tasks necessary to complete an entire job, such as setting up work schedules and making assignments to team members.

Cross-functional teams represent an attempt to modify the classic hierarchical form of an organization based on a vertical chain of command. They include horizontal coordination in order to plan and control processes that flow laterally. If no lateral coordination is achieved, the organization becomes a collection of islands of specialization, without integration of business processes that flow horizontally across the organizational chart. The concept of linking cross-functional processes was shown in Figure 3-1. Note that a cross-functional approach achieves the objectives of customer, functions, processes, and the total organization.

TRAINING and DEVELOPMENT

Increased involvement means more responsibility, which in turn requires a greater level of skill. This must be achieved through training. Baldrige Award winners place a great deal of emphasis on training and support it with appropriate provision of resources. Motorola allocates about 2.5 percent of payroll costs or $120 million annually to training, 40 percent of which goes to quality training. The company calculates the training return at about $29 for each $1 invested. Additional benefits include (1) improved communications, (2) change in corporate culture, and (3) demonstration of management's commitment to quality. (Xerox has extended quality training to 30,000 supplier personnel.)

▪ Since the early 1980s, Hughes Aircraft has made quality one of its chief operating philosophies. The cornerstone of the company's TQM thrust is continuous measurable improvement (CMI). Recently, the firm has championed a unique "trickle-down" training system to sustain its quality and productivity improvements. Under CMI (Cascaded Training Program), the managers responsible for achieving improvement teach the philosophy and principles of CMI leadership throughout the organization.[2]

Although the type of training depends on the needs of the particular company and may or may not extend to technical areas, the one area that should be common to all organization training programs is *problem solving*. Problem solving should be institutionalized and internalized in many, if not most, companies. This would be a prerequisite to widespread empowerment.

Training usually falls into one of three categories: (1) reinforcement of the quality message[3] and basic skill remediation, (2) job skill requirements, and (3) knowledge about principles of TQM. The latter typically covers problem-solving techniques, problem analysis, statistical process control, and quality measurement—areas that go beyond typical job skills. If groups or teams are utilized, training in the group process and group decision making is included. According to a survey conducted by the Conference Board, top companies commonly address the following topics in quality training curricula:

- Quality awareness
- Quality measurement (performance measures/quality cost benchmarking, data analysis)
- Process management and defect prevention
- Team building and quality circle training
- Focus on customers and markets
- Statistics and statistical methods
- Taguchi methods

> ■ Research Testing Laboratories, Inc., a TQM company providing clinical research services, encourages employees to make changes in processes in order to minimize and eliminate errors early in the work process. The goal is 100 percent customer satisfaction. To achieve this goal, employees are provided with a 25-hour training program in which they learn (1) effective interactive skills, (2) the problem-solving process, and (3) the quality improvement process.

Managerial training may take the form of item 3 above (TQM principles). In addition, programs often are directed toward sensitizing individuals to the strategic importance of quality, the cost of poor quality, and their role in influencing the quality of products and services.

The International Quality Study (IQS) was conducted among 584 companies representing four industries. The use of quality tools in the American auto industry is expected to increase 1.5- to 6-fold over the next three years. Quality training was found to have the greatest impact when coupled with other practices, such as measurement and reward systems.[4]

SELECTION

Selection is choosing from a group of potential employees (or placement from existing employees) the specific person to perform a given job. In theory, the process is simple: decide what the job involves and what abilities are necessary, and then use established selection techniques (ability tests, personality tests, interviews, assessment centers) as indicators of how the candidate will perform.

The process is not so simple, however, when TQM enters the picture. The job requirements for a typist, a machinist, or even a manager can be determined by job analysis, and the qualifications of a candidate can be compared to these requirements. When a company commits to TQM, an entirely new dimension is introduced. The skills and abilities required for a specific job can usually easily be identified and then matched with an individual. People well suited for operating in a quality climate may require additional characteristics, such as attitude, values, personality type, analytical ability.

Persons working in a quality environment need sharp problem-solving ability in order to perform the quantitative work demanded by statistical process control, Pareto analysis, etc. Because of the emphasis on teams and group process, personnel must function well in group settings. Motorola shows applicants video tapes of problem-solving groups in action and asks them how they would respond to a particular quality issue. Presumably this technique encourages *self-selection.*

What is perhaps different in the selection process in a TQM environment is the emphasis on a *quality-oriented organization culture* as the desired outcome of the selection process.[5]

PERFORMANCE APPRAISAL

The purpose of performance appraisal is to serve as a diagnostic tool and review process for development of the individual, team, and organization. Appraisals are used to determine reward levels, validate tests, aid career development, improve communication, and facilitate understanding of job duties.[6]

Deming cites *traditional* employee evaluation systems as one of seven deadly diseases confronting U.S. industry. He states that *individual* performance evaluations encourage short-term goals rather than long-term planning. They undermine teamwork and encourage competition among people for the same rewards. Moreover, the overwhelming cause of non-quality is not the employee but the system; by focusing on individuals, attention is diverted from the root cause of poor quality: the system.

Many TQM proponents, like Deming, argue that traditional performance appraisal methods are attempts by management to pin the blame for poor organization performance on lower level employees, rather than focusing attention on the system, for which upper management is primarily responsible.

Should individual performance appraisal be eliminated, as Deming suggests?[7] This is unlikely in view of the historical and widespread use of this human resource management tool. What, then, can be done to relate individual and group performance to a total quality strategy?

Performance appraisals are most effective when they focus on the objectives of the company and therefore of the individual or group. Because the eventual outcome of all work is quality and customer satisfaction, it follows that appraisal should somehow relate to this outcome—to the objectives of the company, the group, and the individual. In other words, a performance appraisal system should be aligned with the principle of shared responsibility for quality. This can be accomplished by focusing on development of the skills and abilities necessary to perform well and, as such, directly support collective responsibility.

■ In a model used by the Hay Group (a consulting organization), individuals are evaluated for base pay on such variables as ability to communicate, customer focus, and ability to work as a team. Managers are rated on employee development, group productivity, and leadership. Variable pay for both is based on what is accomplished. Because customer focus is a critical part of any TQM effort, a three-category rating system that involves (1) not meeting customer expectations, (2) meeting them, and (3) far exceeding them is easy to implement.[8]

Answering Deming and the other critics is not easy. The integration of total quality and performance appraisal is necessary. One should reinforce the other. One approach might be to modify existing systems in accordance with the following principles:

■ Customer expectations, not the job description, generate the individual's job expectation.

■ Results expectations meet different criteria than management-by-objectives statements.

■ Performance expectations include behavioral skills that make the real difference in achieving quality performance and total customer satisfaction.

■ The rating scale reflects actual performance, not a "grading curve."

■ Employees are active participants in the process, not merely "drawn in."

Regardless of which specific system is adopted, there seems to be little question that performance management practices need to be in line with and supportive of TQM.

COMPENSATION SYSTEMS

This may be one of the most elusive and controversial of all systems that support TQM. Historically, compensation systems have been based on (1) pay for performance or (2) pay for responsibility (a job description). Each of these is based on individual performance, which creates a competitive atmosphere among employees. In contrast, the TQM philosophy emphasizes flexibility, lateral communication, group effectiveness, and responsibility for an entire process that has the ultimate outcome of customer satisfaction. No wonder research and writing have offered little in the way of new approaches that are more in tune with the needs of TQM.

■ Shawnee Mission (Kansas) Medical Center attempted to set up an infrastructure to push TQM ideals throughout the organization. In 1992 the center operationalized its new evaluation system based on personal development, education, and teamwork. Everyone receives the same raise.

Both training and performance appraisal are desirable components of a TQM implementation strategy, but compensation is an equally necessary dimension. Employees may perceive the system as a reflection of the company's commitment to quality.

Individual or Team Compensation?

A company's infrastructure, specifically its reward and compensation systems, provides an accurate picture of its strategic goals. If compensation criteria are focused exclusively on individual performance, a company will find that initiatives promoting teamwork may fail. A TQM vision and the principles supporting it are unlikely to take hold unless the values on which they are based are built into the underlying structure.

■ Target Stores is among the growing number of companies in the retail industry that are going beyond logistics-specific performance measures and are tying pay into the effectiveness of TQM programs. Throughout the logistics field, pay for performance and pay for quality appear to becoming more entrenched.

There is no lack of compensation plans in U.S. industry. Gain sharing, profit sharing, and stock ownership are among the systems designed to create a financial incentive for employees to be involved in performance improvements. Gain sharing is one of the most rapidly growing compensation and involvement systems in U.S. industry. It is a system of management in which an organization seeks higher levels of performance through the involvement and participation of its people. Employees share financially in the gain when performance improves. The approach is a team effort in which employees are eligible for bonuses at regular intervals on an operational basis. Gain sharing reinforces TQM, partially because it contains common components, such as involvement and commitment.[9]

The jury is still out on the effectiveness of these plans, but evidence suggests that effectiveness is a function of strong communication programs and widespread employee involvement.

Summary

Many reasons have been offered as the cause of poor performance in organizations: (1) system failure; (2) misunderstanding of job expectations; (3) lack of awareness about performance; (4) lack of time, tools, or resources to succeed; (5) lack of necessary knowledge or skills; (6) lack of appropriate consequences for performance; and (7) bad fit for the job. Although a compensation system supportive of TQM is not the only remedy, combined with other human resource management systems it will go a long way toward improvement of performance and development among individuals, groups, and the organization.

TOTAL QUALITY ORIENTED
HUMAN RESOURCE MANAGEMENT

Human resource executives are faced with both a challenge and an opportunity. They are not generally perceived with the same regard as line managers. Philip Crosby describes the human resource department as behind the times and the human resource executive as his or her own worst enemy. On the other hand, the department can play a critical role in the implementation of a holistic quality environment in support of a strategic initiative. To accomplish this role, the function should not only be designed to support TQM throughout the organization, but should make sure that good quality management practices are followed within the processes of the function itself. This means continuous improvement as a way of department life. Bowen and Lawler suggest putting the following principles of TQM to work *within* the human resource department:[10]

1. Quality work the first time
2. Focus on the customer
3. Strategic holistic approach to improvement
4. Continuous improvement as a way of life
5. Mutual respect and teamwork

It is evident that some modification of traditional human resource management practices is required if the function is to support the TQM program throughout the company. Planning is the first step. The 1993 Baldrige Award criteria describe human resource planning:[11]

> Human resource plans might include the following: mechanisms for promoting cooperation such as internal customer/supplier techniques or other internal partnerships; initiatives to promote labor–management cooperation, such as partnerships with unions; creation and/or modification of recognition systems; mechanisms for increasing or broadening employee responsibilities; creating opportunities for employees to learn and use skills that go beyond current job assignments through redesign of processes; creation of high performance work teams; and education and training initiatives. Plans might also include forming partnerships with educational institutions to develop employees or to help ensure the future supply of well-prepared employees.

QUESTIONS for DISCUSSION

5-1 Would a quality improvement program based on process control be more appropriate for employee involvement than a system based on traditional production methods? If so, explain why.

5-2 What effect does employee involvement have on motivation? Explain the effect in terms of motivational theory.

5-3 Contrast the benefits of the different types of small groups or teams. Which would be more appropriate for achieving integration across organizational functions or departments?

5-4 A Deming principle advises to "create conditions that allow employees to have pride in their workmanship." What are these conditions and how can they be implemented?

5-5 Assume that a company has just committed to change from a traditional style of management to one based on TQM. What topics would you include for

■ Shop floor employees

■ Front-line supervisors

■ Middle-level managers

5-6 Describe how training in problem solving would improve

■ Process control

■ Employee motivation

ENDNOTES

1. *HR Focus,* Jan. 1993, pp. 1, 4.
2. Judy Rice, "Cascaded Training at Hughes Aircraft Helps Ensure Continuous Measurable Improvement," *National Productivity Review,* Winter 1992/1993, pp. 111–116.
3. Bernie Knill, "The Nitty-Gritty of Quality Manufacturing," *Materials Handling Engineering,* July 1992, pp. 40–42. In a Conference Board survey, training is first used to reinforce the quality message and then to build skills. Another finding of the survey is that leaders link TQM to performance review and compensation.
4. Trace E. Benson, "When Less Is More," *Industry Week,* Sep. 7, 1992, pp. 68–77.

5. David E. Bowen and Edward E. Lawler III, "Total Quality-Oriented Human Resource Management," *Organization Dynamics,* Spring 1992, pp. 29–41.
6. David E. Bowen and Edward E. Lawler III, "Total Quality-Oriented Human Resource Management," *Organization Dynamics,* Spring 1992, p. 36.
7. Some recent articles that treat performance appraisal in a TQM context include Kathleen A. Guinn, "Successfully Integrating Total Quality Management and Performance Appraisal," *Human Resource Professional,* Spring 1992, pp. 19–25; Mike Deblieux, "Performance Reviews Support the Quest for Quality," *HR Focus,* Nov. 1991, pp. 3–4; Jean B. Ferketish and John W. Hayden, "HRD & Quality: The Chicken or the Egg?" Jan. 1992, pp. 38–42.
8. Linda Thornburg, "Pay for Performance: What You Should Know (Part 1)," *HR Magazine,* June 1992, pp. 58–61.
9. Robert L. Masternak, "Gainsharing at B. F. Goodrich: Succeeding Together Achieves Rewards," *Tapping the Network Journal,* Fall/Winter 1991, pp. 13–16.
10. David Bowen and Edward Lawler, "Total Quality-Oriented Human Resources Management," *Organizational Dynamics,* Spring 1992, p. 29.
11. *Malcolm Baldrige National Quality Award. 1993 Award Criteria,* Gaithersburg, Md.: U.S. Department of Commerce, National Institute of Standards and Technology, 1993, p. 21.

6

MANAGEMENT of PROCESS QUALITY

A Deming-style "total quality management" approach to improving service quality is rooted in the unglamorous and never fashionable discipline of statistics. Using Mr. Deming's statistical approach to total quality management, we have reduced service expenses 35% over the past 12 months while improving service quality.

President
Savin Copiers

The need for top management to display leadership in setting the climate and culture for total quality management was outlined in Chapter 2. Climate and culture, however, are not enough. It is unlikely that exhortations and slogans will be effective unless accompanied by action planning and implementation. A statement such as "We Are the Quality Company" convinces no one—not the employees and not the customers. The company should be organized for quality assurance in the context of modern quality management.

Assume that the criteria of the Baldrige Award fairly represent what is generally accepted as the national standard for management of process quality:

■ The *Management of Process Quality* category examines systematic processes the company uses to pursue ever-higher quality and company operational performance. The key elements of process management are examined, including research and development, design, management of process quality for all work units and suppliers, systematic quality improvement, and quality assessment.

It is apparent that this definition is directly related to how well the *processes* are managed—*all* of the processes in the organization that contribute directly or indirectly to quality as the customer defines it. The concept is illustrated in Figure 6-1. Note that the control component (quality assurance) has moved from measuring output (the traditional control system) to controlling the *continuous improvement of the process.*

The traditional approach to quality control was inspection of the final product, and this approach is still practiced by many firms. This chapter will introduce methods and techniques that are significantly more advanced and more effective than the practice of "final inspection," which has been used for so long. Although the concepts in this chapter are not the last word in modern total quality management (TQM), they represent substantial potential for improving quality, cost, and productivity in almost any company.

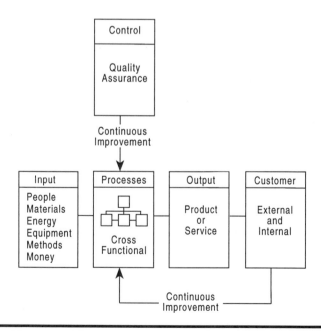

Figure 6-1 Management System

A BRIEF HISTORY of QUALITY CONTROL

Concern for product quality and process control is nothing new. Historians have traced the concept as far back as 3000 B.C. in Babylonia. Among the references to quality from the Code of Hammurabi, ruler of Babylonia, is the following excerpt: "The mason who builds a house which falls down and kills the inmate shall be put to death." This law reflects a concern for quality in antiquity.[1] Process control is a concept that may have begun with the pyramids of Egypt, when a system of standards for quarrying and dressing of stone was designed. One has only to examine the pyramids at Cheops to appreciate this remarkable achievement. Later, Greek architecture would surpass Egyptian architecture in the area of military applications. Centuries later, the ship-building operations in Venice introduced rudimentary production control and standardization.

Following the Industrial Revolution and the resulting factory system, quality and process control began to take on some of the characteristics that we know today. Specialization of labor in the factory demanded it. Interchangeability of parts was introduced by Eli Whitney when he manufactured 15,000 muskets for the federal government. This event was representative of the emerging era of mass production, when inspection by a skilled craftsman at a workbench was replaced by the specialized function of inspection conducted by individuals not directly involved in the production process.

Specialization of labor and quality assurance took a giant step forward in 1911 with the publication of Frederick W. Taylor's book *Principles of Scientific Management*.[2] This pioneering work had a profound effect on management thought and practice. Taylor's philosophy was one of extreme functional specialization and he suggested eight functional bosses for the shop floor, one of whom was assigned the task of inspection:

> The inspector is responsible for the quality of the work, and both the workmen and the speed bosses [who see that the proper cutting tools are used, that the work is properly driven, and that cuts are started in the right part of the piece] must see that the work is finished to suit him. This man can, of course, do his work best if he is a master of the art of finishing work both well and quickly.[3]

Taylor later conceded that extreme functional specialization has its disadvantages, but his notion of process analysis and quality control by inspection of the final product still lives on in many firms today. Statistical quality control (SQC), the forerunner of today's TQM or total quality control, had its beginnings in the mid-1920s at the Western Electric plant of the Bell System. Walter Shewhart, a Bell Laboratories physicist, designed the original version of SQC for the zero defects mass production of complex telephone exchanges and telephone sets. In 1931 Shewhart published his landmark book *Economic Control of Quality of Manufactured Product.*[4] This book provided a precise and measurable definition of quality control and developed statistical techniques for evaluating production and improving quality. During World War II, W. Edwards Deming and Joseph Juran, both former members of Shewhart's group, separately developed the versions used today.

It is generally accepted today that the Japanese owe their product leadership partly to adopting the precepts of Deming and Juran. According to Peter Drucker, U.S. industry ignored their contributions for 40 years and is only now converting to SQC.[5]

▪ The Willimatic Division of Rogers Corporation, an IBM supplier, uses just-in-time techniques along with X-bar and R charts for key product attributes to achieve statistical process control. Rework is reduced by 40 percent, scrap by 50 percent, and productivity is increased by 14 percent.[6]

PRODUCT INSPECTION vs. PROCESS CONTROL

Structure follows strategy.

Nothing happens until a sale is made.

If you can't measure it, you can't manage it.

These statements are typical of the popular catchphrases adopted by particular functions (e.g., planning, sales, accounting) within the business. The popularity of the expression usually means that there is a measure of truth behind it. Truisms in the field of quality management include "don't inspect the product, inspect the process" and "you can't inspect it in, you've got to build it in."

There is sound thinking behind these two statements. In the previous discussion of the control process, the point was made that controlling the output of the system *after the fact* was historical action and nothing could be done to correct the variation after it had already occurred. This is feedback control. The same is true of inspecting the product. The variation or the defect has already occurred. What is needed is a feedforward system that will prevent defects and variations. Better yet is a system that will improve the process. This is the idea behind process control (Figure 6-1).

What is the process? Does it begin with material inspection at the receiving dock and end with final inspection, or does it begin with design and end with delivery to the customer? Does it begin with market research and end with after-sale service? If we take the broader view, the process might begin with the concept of the product idea and extend through the life cycle of the product to ultimate maturity and phase out. This definition matches the concept of TQM.

It is clear that in the philosophy of TQM, most (if not all) business functions and activities (i.e., processes) are interrelated and none stand alone—not purchasing, engineering, shipping, order processing, or manufacturing. Key business objectives and organization success are dependent on cross-functional processes. Moreover, these processes must change as environments change. The conclusion emerges that true process optimization requires the application of tools and methods in all activities, not just manufacturing.

Historically, there have been two major barriers to effective process control. The first has been the tendency to focus on volume of output rather than quality of output. Volume of production has been the major objective in the mistaken notion that more units of output means lower unit cost. Another barrier is the quality control system that measures products or service against a set of internal conformance specifications that may or may not relate to customer expectations. The result in many cases has been inferior quality products that are reworked or scrapped or, worse, products that customers *did not buy.* As will be discussed in Chapter 11, the cost of poor quality can amount to 25 to 30 percent of sales revenues. The profit potential in quality improvement is greater than simply improved production of inferior quality.

▪ Bytex Corporation manufactures electronic matrix switches for Citicorp, MasterCard, American Express, and others. The company has focused on understanding the process, concentrating on eliminating non-value-added transactions. Cycle time is down by 60 percent, inventory down by 43 percent, final assembly time down by 52 percent, and floor space down by 30 percent. The resulting product is superb.[7]

MOVING from INSPECTION to PROCESS CONTROL

Process control may still require measurement that is determined by inspection, but the activity of inspection is now transformed into a diagnostic role. The objective is not merely to discover defects, but rather to identify and remove the cause(s) of defects or variations. Process control now becomes problem solving for *continuous improvement*. Moving from inspection to process control takes place in steps or phases:

Step	Action
1	Process characterization Definition of process requirements and identification of key variables
2	Develop standards and measures of output Involve work force
3	Monitor compliance to standards and review for better control Identify any additional variables that affect quality
4	Identify and remove cause(s) of defects or variations (this requires a step-by-step documentation of the process and process control charting)
5	Achievement of process control with improved stability and reduced variation

STATISTICAL QUALITY CONTROL

This is the oldest and most widely known of the several process control methods. It involves the use of statistical techniques, such as control charts, to analyze a work process or its outputs. The data can be used to identify variations and to take appropriate actions in order to achieve and maintain a state of statistical control (predetermined upper and lower limits) and to improve the capability of the process. It is the best-known innovation among Deming's ideas.

Rigorously applied, SQC can virtually eliminate the production of defective parts.[8] By identifying the quality that can be expected from a given production process, control can be built into the process itself. Moreover, the method can spot the causes of variations—incoming materials, machine calibration, temperature of soldering iron, or whatever.

Despite the maturity of the method and its proven benefit, many firms do not take full advantage of it. One survey found that 49 percent of responding electronic manufacturers reported using SQC techniques, but 75 percent of them also continued to use traditional 100 percent inspection. This is in an industry where quality in the manufacturing process is essential.

▪ At Motorola, SQC has been integrated into the corporate culture and is being applied in all areas of the plant. Steps to place a process under statistical control include (1) characterizing the process, (2) controlling it, and (3) adjusting the process when non-random deviations are observed. Six sigma is the goal.

The term *statistical process control* can be misleading because it is so frequently confined to manufacturing processes, whereas the methods can be useful for improving results in other non-manufacturing areas such as sales and staff activities. Moreover, the methods can be used in many of the activities and functions of service industries. It is also worth noting that the only universal technique for SQC is logical reasoning applied to the improvement of a process. Thus it is a systematic way of problem solving.

A *process* is a set of causes and conditions and a set of steps comprising an activity that transforms inputs into outputs. Consider the number of processes involved in the airline industry: the process of taking and confirming a reservation, of baggage handling, of loading passengers, of meal service, etc. The process is any set of people, equipment, procedures, and conditions that work together to produce a result—an output.

The process is expected to add value to the inputs in order to produce an output. The ratio of output to input is called productivity and the objectives are to (1) increase the ratio of output to input and (2) reduce the variation in the output of the process. If the variation is too small or insignificant to have any effect on the usefulness of the product or service, the output is said to be within tolerance. Should the output fall outside the desired tolerance, the process can be improved and returned to tolerance by defining the cause of the change (the problem) and taking action to make sure that the cause does not recur.

BASIC APPROACH to
STATISTICAL QUALITY CONTROL[9]

SQC and its companion *statistical process control* (SPC) were developed in the United States in the 1930s and 1940s by W. A. Shewhart,

W. E. Deming, J. M. Juran, and others. These techniques (some call them philosophies) have been used for decades by some American firms and many Japanese companies. Despite the proven effectiveness of the techniques, many U.S. firms are reluctant to use them.[10]

The approach is designed to identify underlying cause of problems which cause process variations that are outside predetermined tolerances and to implement controls to fix the problem. The basic approach contains the following steps:

1. Awareness that a problem exists.
2. Determine the specific problem to be solved.
3. Diagnose the causes of the problem.
4. Determine and implement remedies to solve the problem.
5. Implement controls to hold the gains achieved by solving the problem.

MANUFACTURING to SPECIFICATION vs. MANUFACTURING to REDUCE VARIATIONS

Among production managers who manufacture to specifications or those who depend upon final inspection, the common problem can be traced to the control loop. Defect statistics are generated by inspection, but appropriate action is not taken to define problems, determine cause(s), and correct variations. Companies continue to live with a reject rate that is considered to be "normal," as typified by statements such as "We can live with X percent defectives" or "that's fairly common in the industry."

The benefit of manufacturing to reduce variations (process control) is generally recognized.[11] It is the purpose of SQC to *identify* and *reduce* variations from standard and *continuously improve* the process until a theoretical condition of "zero defects" is achieved.[12] The causes of variations are many and vary from industry to industry. Common sources include (1) material balance disturbances, (2) energy balance changes, (3) process instabilities, (4) equipment failure and wear, and (5) poor control loop performance.[13] SQC is used to develop control limits for each step within the process. Measuring sample parts and graphing trends leads to identification of the cause(s) of any erratic (non-random) behavior in the process.

The objective of process control is not only production of quality output, but reduction of costs as well. Quality is defined as the total acceptable variation divided by the total actual variation or Cp index. When used alone, this measure may be misleading because it assumes acceptable quality product design.[14] This, of course, is not always the case and suggests the need for the cross-functional process control mentioned earlier.

Data acquisition and monitoring is an essential step if the process is to remain in control. This tracking is generally accomplished by the operator concerned. In more sophisticated plants, particularly in unattended manufacturing, the goal is to have in-process measurement and correction in real time through the use of sensors or other measuring devices.[15] Devices such as bar code readers, vision systems, and counters are some of the tools available for collection of data. Of course, data alone is not enough. Data must be organized in such a way that process decisions can be made.

PROCESS CONTROL in SERVICE INDUSTRIES

Examination of the U.S. Government Standards Industrial Classification of Industries suggests many industries in which the use of SPC would be appropriate. Use of the techniques is spreading to such industries as transportation,[16] health care, and banking.[17]

To some extent the service process is more difficult to control than manufacturing because quality is typically measured at the customer interface, when it is already too late to fix the problem. Hence, "final inspection" will always be a part of the process; the customer serves as the inspector.

Service failures are analogous to bad parts in manufacturing, and measures of service may be compared to manufacturing tolerances or standards. SPC can be used to measure consistency of service and determine causes of deterioration from prescribed standards and the cause(s) of variations. In transportation the cause may be missed appointments, refusals, or weekend closures.[18] At the First National Bank of Chicago, a number of processes are checked weekly against over 500 customer-sensitive measures.[19]

■ L. L. Bean, a mail order company in Freeport, Maine, is known worldwide for its outstanding distribution system. It is the ideal company to benchmark for that function. The company analyzed all key activities and processes, including benchmarking competitors. It is ranked number 1 in virtually every product category in which it is evaluated by outside sources.[20]

Customer Defections: The Measure of Service Process Quality

Measures of output, as the customer defines them, are not too difficult to identify in service firms. An airline can measure on-time departures and the time it takes to make a reservation. A bank can measure the ratio of ATM downtime to total number of ATM minutes available and so on. Measures such as these are necessary, but the most important measure is *customer defections* or customers lost to the competition.

What is the cost of a customer defection? Conversely, what is the value of a customer retention? Defections have a substantial effect on profits and cost, more so than market share, economies of scale, or unit costs. Simply put, losing a customer costs money and retaining one makes money.

The initial cost of acquiring a new customer involves a number of one-time costs for prospecting, advertising, records, and such. Banks, attorneys, mutual funds, and credit card companies are examples of firms that spend to recruit a customer and establish an account. However, once a relationship is established, the marginal cost of each additional dollar of sales *diminishes*—provided the customer does not defect.

Improving the processes and reducing the process variations that reduce customer defections can be perceived not as a cost but as an investment. Consider the following examples:

■ Taco Bell calculates that the lifetime value of a retained customer is $11,000.

■ An auto dealer believes that the lifetime value of retaining a customer is $300,000 in sales.

■ MBNA America, a credit card company, has found that a 5 percent improvement in defection rates increases its average customer value by 125 percent.

PROCESS CONTROL for INTERNAL SERVICES

■ Until it moved to Raleigh, North Carolina, IBM's personal computer assembly operation was located at its plant in Boca Raton, Florida. The general manager was committed to internal as well as external quality. In support of this commitment, the following policy was adopted, widely disseminated, and implemented through "Excellent Plus" groups:

Excellence Plus Commitment

IBM Boca Raton will deliver defect-free, competitive products and services, on time, to all customers. Quality will be the primary consideration in all decisions related to cost and delivery. *Likewise, each department will provide defect-free work to the next user of its output or service* (italics added).

An inventory of the many functions and activities in an organization will reveal that each activity is responsible for the operations of one or more processes where the customer is an *internal* user of its output or service. Many, if not most, of these processes lend themselves to process control methods.

AT&T's support services organization in Chicago is responsible for word processing and reprographics. Through SPC, a fivefold improvement in typing accuracy and a halving of turnaround time in reprographics was achieved. Most of the gain was attributed to better communications with customers.[21]

QUALITY FUNCTION DEPLOYMENT

For centuries, and even today, navies have built ships in the same process sequence:

Design → Build hull and launch → Outfit →

Trial run → Return to shipyard → Rework →

Operational check → Return → Fix → Operational

This sequence in modern construction of ships and other weapon systems almost always results in time and cost overruns and subsequent operational deficiencies. This is evidence of inadequate process control, which may change as a result of the Defense Department's shift from testing the product to testing the process. This shift is a part of the Pentagon's recent TQM strategy.[22]

It is generally agreed that maybe nine out of ten new product developments end up as a design, manufacturing, or marketing failure. These failures may be more the fault of the organization than the market. Many firms lack a system to integrate the market demands with the organization processes. Most applicants for the Baldrige Award, according to examiners, lack management processes that ensure the efficient flow of customer demands throughout the organization.[23]

If quality definition (customer expectation) is not introduced early in the concept or design stage, there is the risk (indeed the probability) that design errors and product defects will only be discovered at later stages of production or final inspection. The worst scenario is discovery by the customer in the marketplace. Motorola estimates that whereas design accounts for only 5 percent of product cost, it accounts for 70 percent of the influence on manufacturing cost.

The major functions of the organization and the matching activities/ processes are shown in Figure 6-2. Each is necessary throughout the life cycle of the product, but if the beginning of each process or activity must wait for the end of the preceding one, the time to market is lengthened and the product may be obsolete or overtaken by competition midway through the processes. A method is needed to integrate all processes and relate them to the customer.

Every chief executive officer would welcome a TQM system that would:

■ Implement strategic quality management, including market segment differentiation based on customer expectations

■ Communicate a culture of quality throughout the organization

■ Translate technical requirements into process requirements and then to production planning

■ Organize the potential for world-class competition

■ *Integrate*
1. The special interest functions of the company
2. The stream of processes and provide a basis for process design and control
3. Suppliers and customers
4. Everyone in the process while promoting a team culture with interfunctional teams

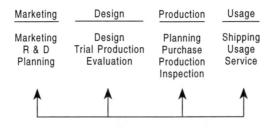

Marketing	Design	Production	Usage
Marketing R & D Planning	Design Trial Production Evaluation	Planning Purchase Production Inspection	Shipping Usage Service

Figure 6-2   Quality Function Deployment Chart

This is a lot to ask of any method, but proponents of quality function deployment (QFD) suggest that this method has the potential to achieve many of these requirements. It has proven so effective as a competitive advantage in some companies (e.g., Ford, Digital Equipment, Black & Decker, Budd, Kelsey Hayes) that they are unwilling to talk about it.[24] In Japan, where the method was first used, companies have achieved dramatic improvement in the design-development process, including reductions of 30 to 50 percent in engineering changes and design-cycle time and 20 to 60 percent in start-up costs.[25]

QFD is a group of techniques for planning and communicating that coordinates the activities within an organization. It is a dynamic, iterative method performed by interfunctional teams from marketing, design, engineering, manufacturing engineering, manufacturing, quality, purchasing, and accounting and in some cases suppliers and customers as well. Thus, a common quality focus is achieved across all functions: quality function deployment. The basic premise is that products should be designed to reflect the desires and tastes of customers. An additional benefit is improvement of the company's management processes.[26]

The primary technique is a visual planning matrix called the "House of Quality" which links customer requirements, design requirements, target values, and competitive performance in one easy-to-read chart. The concept, but not the details, is illustrated in Figure 6-3.[27]

QFD unfolds in the following steps or phases. Note that step numbers for product planning and design processes are entered in the sections of the House of Quality in Figure 6-3.

Step 1   **Product planning.**   This begins with customer requirements, defined by specific and detailed phrases that the customers

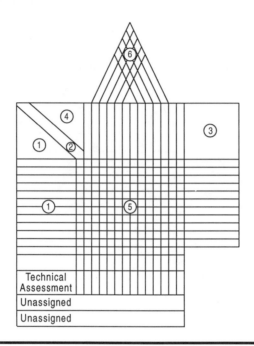

Figure 6-3 House of Quality Concept

in their own words use to describe desired product characteristics.

▪ Eaton Corporation, a supplier to the automobile industry, selected a control device for a QFD pilot process. A
matrix chart was prepared that related desired product
features to part quality characteristics. Each quality characteristic was ranked. Through QFD, selling price and engineering expenses were reduced by 50 percent.[28]

Step 2 **Prioritize** and **weight** the relative importance that customers
have assigned to each characteristic. This can be done on a
scale (e.g., 1 to 5) or in terms of percentages that sum to 100
percent.[29]

Step 3 **Competitive evaluation.** For those who want to be world-
class or meet or beat the competition, it is essential to know
how their products compare. Specifically, will the characteristics identified in steps 1 and 2 provide a strategic competitive advantage? (See Chapter 8 for further information on
benchmarking.)

Step 4 **The design process.** This is where the customer's product characteristics meet the *measurable* engineering characteristics that directly affect customer perceptions.

Step 5 **Design** (continued). The central relationship matrix indicates the degree to which each engineering characteristic affects the customer's characteristics. Strengths of relationships are entered.

Step 6 **Design** (continued). The roof of the "house" matrix encourages creativity by allowing changes between steps 4 and 5 in order to judge potential trade-offs between engineering and customer characteristics.

Step 7 **Process planning.** Output from the design process goes to process planning, where the key processes (e.g., cutting, stamping, welding, painting, assembly, etc.) are determined. This step may have its own matrix.

Step 8 **Process control.** Output from step 7 goes to process control, where the necessary process flows and controls are designed.

The entire QFD process is "deployed" as illustrated in Figure 6-4. The "hows" of one step become the "what's" of the next. Many of the statistical techniques mentioned previously can be used. Market research has particular methods for that function.

In all cases the interfunctional teams are involved. This is necessary to avoid rework and redesign as well as overruns in cost and time. Questions need to be answered along the way: "What does the customer really want?" "Can we design it?" "Can we make it?" "Is it competitive?" "Can we sell it at a profit?" "Do the processes support it?" Hewlett-Packard estimates that quality programs have saved the company $400 million in warranty costs. Prior to implementing QFD and quality programs, the company estimated that non-quality costs added up to 25 to 30 percent of sales dollars.[30]

The essential prerequisite for QFD is the determination of customer requirements that are defined by specific and detailed phrases that customers in their own words use to describe desired product characteristics. To achieve this degree of specificity, it may be necessary to communicate with customers one-on-one or in focus groups. A less desirable method is to use surveys or other means.

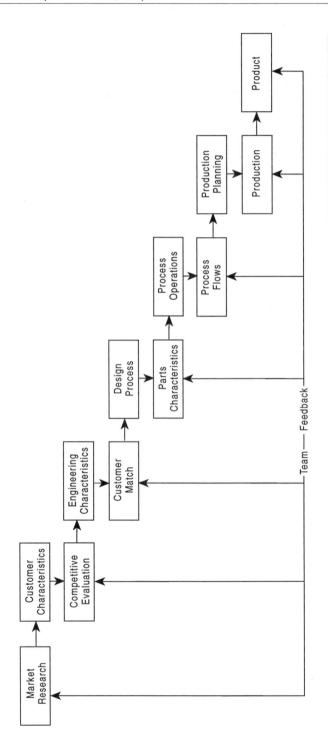

Figure 6-4 Quality Function Deployment

JUST-in-TIME (JIT)

■ By the third year of JIT implementation, Isuzu (a Japanese company) had reduced the number of employees from 15,000 to 9,900, reduced work-in-process from 35 billion yen to 11 billion yen, and decreased the defect rate by two-thirds.[31]

■ Hewlett-Packard has spread JIT to all areas, including cost accounting, procurement, and engineering. At one plant where 290 pieces of equipment are hand assembled, product reliability has improved sixfold and productivity is up considerably.[32]

■ As part of its conversion to JIT, Westinghouse Electric's Asheville, North Carolina plant was run as a number of mini plants. Cycle time has been reduced two to four times, on-time performance is up over 90 percent, and shop productivity is up by 70 percent. Employees are trained to perform multiple functions, and each will end up knowing how to build the complete product.

U.S. manufacturing has been characterized by mass production, high-volume output, and machine capacities that are pushed to the limit. This is changing as American managers begin to discover a production method called just-in-time (JIT). Proponents say that it is more than a manufacturing system; they call it a philosophy and a way of approaching business goals that incorporates (1) producing what is needed when it is needed, (2) minimizing problems, and (3) eliminating production processes that make safety stocks necessary.[33]

Prior to the 1960s, the goal of production planning was cost optimization. In the early 1970s it became requirements planning, and the technique of materials requirements planning (MRP) computed material needs to meet a sales forecast and production plan. MRP was, and is, an effort to balance the sometimes conflicting demands of safety stocks, inventory carrying costs, economic order quantity, and risk factors related to possible stockouts. Today, the modern corporation is turning to manufacturing as a crucial strategic resource and is adopting JIT as a basic component of manufacturing strategy. The view is that the expense and risk of maintaining inventory can be reduced so that lower costs becomes a way of improving both productivity and quality. Of course, inventory is not the only consideration of JIT. It involves all functions and all processes.

JUST-in-TIME or JUST-in-CASE

JIT infers that "less is best," while just-in-case (JIC) involves the use of buffer or safety stocks. Conventional reasons given to explain the need for buffer stock include avoiding risks of stockouts or failure of suppliers, getting a better price for volume purchasing, or avoiding an anticipated price increase. The presence of such "excess" inventory increased the risk of obsolescence and deterioration, increased the need for warehouse and shop floor space, and by "pushing" parts through the assembly process encouraged a number of wasteful practices. Operators were unconcerned with workstations other than their own. The attitude became "there's plenty more where that came from." If a defective part was discovered, the tendency was to blame it on a previous operation or assume that it would be corrected later in the process or at the rework area.

Shigeo Shingo, who is credited with designing Toyota's JIT production system, believes that the "push" process used in the United States generates process-yield imbalances and interprocess delays.[34] *Kanban,* as JIT is called in Japan, means "visible record." It is a means of pulling parts through the assembly process; production is initiated only when a worker receives a visible cue that assembly is needed for the next step in the process. The worker orders the product from the previous operation so that it arrives just when needed. If one of the key processes fails to produce a quality part, the production line stops. Individual operators are their own inspectors and are cross-trained for a number of tasks. The system is continuously being fine tuned.

Benefits of JIT[35]

JIT is not just an inventory control method. It is a system of factory production that interrelates with all functions and activities. The benefits include:

■ Reduction of direct and indirect labor by eliminating extraneous activities

■ Reduction of floor space and warehouse space per unit of output

■ Reduction of setup time and schedule delays as the factory becomes a continuous production process

■ Reduction of waste, rejects, and rework by detecting errors at the source

■ Reduction of lead time due to small lot sizes, so that downstream work centers provide feedback on quality problems

■ Better utilization of machines and facilities

■ Better relations with suppliers

■ Better plant layout

■ Better integration of and communication between functions such as marketing, purchasing, design, and production

■ Quality control built into the process

The HUMAN SIDE of PROCESS CONTROL

One study found that a very small percentage of employees could define quality or could relate what their companies were doing to improve it.[36]

The problems of managing streams of processes are both methodological and organizational. Peter Drucker concludes that SQC has its greatest impact on the factory's social organization.[37] The essence of his argument relates to the way that the use of statistical tools in the production process places information and hence accountability in the hands of the machine operator rather than non-operators such as inspectors, expediters, repair crews, and supervisors. Each operator becomes his or her own inspector. Operators "own" the machines, which allows them to spot malfunctions and correct problems. The concept is known as "workstation ownership" at IBM, where each employee is responsible for an entire operation in the production line.[38]

If Drucker is right, as he probably is, the potential exists for significant improvement in quality, cost, and productivity. However, there is a down side. Strict adherence to rigid methods and procedures means that workers and teams may lose the autonomy they previously enjoyed, only to have it replaced by the regimentation necessitated by process control. By their very nature SPC and JIT require a focus on the process as a whole, an environment that may be strange to an operator accustomed to the segmented approach previously in effect.[39]

It is probably almost universally accepted that control of any process rests upon measuring against some standard, measure, benchmark, or target. Yet in many organizations, workers and managers operate with two different sets of goals and in two different cultures. It becomes an "us versus them" split culture. As we move from inspection to process control, it is essential that control measures become the property of the workers. SPC and JIT achieve this. Workers are involved in measures over which they have some control in monitoring continuous improvements. Control of measures alone, however, may not be enough. Understanding of and involvement in the system would enhance job satisfaction, which is a necessary dimension. Moreover, like any process or system, the people with hands-on involvement are a valuable resource for refinement and improvement.

Attention to the human resource dimension provides a basis for significant improvements in job development, job satisfaction, training, and morale. Suggested actions to improve the changes include:

- Like all major change, top management support is essential.
- Change the focus from production volume to quality, from speed to flow, from execution to task design, from performing to learning.
- Invest in training, a necessary prerequisite.

QUESTIONS for DISCUSSION

6-1 Explain the difference between feedforward and feedback (final inspection) control. Why is feedforward more appropriate for TQM?

6-2 What are the steps in moving from a system of final inspection to process control?

6-3 Choose a non-manufacturing (service) process and show how statistical quality control would be appropriate.

6-4 How would a *sequential* approach to product design and introduction result in overruns in time, cost, and quality? How would quality function deployment improve the system?

6-5 Is customer defections a measure of service quality? If so, how can the measure be used to reduce customer defections?

6-6 Explain the benefits of just-in-time (JIT).

ENDNOTES

1. Claude S. George, Jr., *The History of Management Thought*, Englewood Cliffs, N.J.: Prentice-Hall, 1972, p. 10. For an excellent summary of sources that have traced the history of the quality movement, see David A. Garvin, *Managing Quality*, New York: The Free Press, 1988, p. 251.

2. Frederick W. Taylor, *Principles of Scientific Management*, New York: Harper & Row, 1911.

3. Frederick W. Taylor, *Shop Management*, New York: Harper & Row, 1919, p. 101.

4. W. A. Shewhart, *Economic Control of Quality of Manufactured Product*, New York: E. Van Nostrand Company, 1931.

5. Peter Drucker, "The Emerging Theory of Manufacturing," *Harvard Business Review*, May/June 1990, p. 95.

6. Harry W. Kenworthy and Angela George, "Quality and Cost Efficiency Go Hand in Hand," *Quality Progress*, Oct. 1989, pp. 40–41.

7. Barbara Dutton, "Switching to Quality Excellence," *Manufacturing Systems*, March 1990, pp. 51–53.

8. Bob Johnstone, "Prophet with Honor," *Far Eastern Economic Review (Hong Kong)*, Dec. 27, 1990, p. 50. A survey of Japanese automobile parts suppliers showed that 93 percent used SPC in their operations.

9. For a detailed treatment of SQC and SPC, see Kaoru Ishikawa, *Guide to Quality Control*, rev. ed., 1982 (available in the United States from UNIPUB [Tel. 800-274-4888]). See also J. M. Juran, *Quality Control Handbook*, 3rd ed., New York: McGraw-Hill, 1974. For application of process control charts, see such standard texts as J. M. Juran and Frank Gryna, Jr., *Quality Planning and Analysis*, New York: McGraw-Hill, 1980 and E. L. Grant and R. S. Leavenworth, *Statistical Quality Control*, 5th ed., New York: McGraw-Hill, 1980.

10. One research study of over 300 U.S. firms found that less than half believed that they had a state-of-the-art quality control program that includes SQC and SPC and utilizes a computer support system. Joel E. Ross and David Georgoff, "A Survey of Productivity and Quality Issues in Manufacturing: The State of the Industry," *Industrial Management*, Jan./Feb. 1991.

11. Ken Jones, "High Performance Manufacturing: A Break with Tradition," *Industrial Management (Canada)*, June 1988, pp. 30–32.

12. Genichi Taguchi and Don Clausing, "Robust Quality," *Harvard Business Review*, Jan./Feb. 1990, pp. 65–75. This article provides a summary of the collection now known as Taguchi methods, a popular approach that is opposed to the zero defects concept, based on the conclusion that the concept promotes quality in terms of acceptable deviation from targets rather than a consistent effort to hit them. Zero defects, according to Taguchi, fixes design before the effects of the quality program are felt.

13. Kenneth E. Kirby and Charles F. Moore, "Process Control and Quality in the Continuous Process Industries," *Survey of Business*, Summer 1989, pp. 62–66.

14. Larry H. Anderson, "Controlling Process Variation Is Key to Manufacturing Success," *Quality Progress*, Aug. 1990, pp. 91–93.

15. Chester Placek, "CMMs in Automation," *Quality*, March 1990, pp. 28–38.

16. Ray A. Mundy, Russel Passarella, and Jay Morse, "Applying SPC in Service Industries," *Survey of Business,* Spring 1986, pp. 24–29.

17. Aleta Holub, "The Added Value of the Customer–Provider Partnership," in *Making Total Quality Happen,* Research Report No. 937, New York: The Conference Board, 1990, pp. 60–63.

18. John E. Tyworth, Pat Lemons, and Bruce Ferrin, "Improving LTL Delivery Service Quality with Statistical Process Control," *Transportation Journal,* Spring 1989, pp. 4–12.

19. Aleta Holub, Endnote 17.

20. Thomas C. Day, "Value-Driven Business = Long-Term Success," in *Total Quality Performance,* Research Report No. 909, New York: The Conference Board, 1988, pp. 27–29.

21. Laurence C. Seifert, "AT&T's Full-Stream Quality Architecture," in *Total Quality Performance,* Research Report No. 909, New York: The Conference Board, 1988, pp. 47–49.

22. Pam Nazaruk, "Commitment to Quality: Test Process Not Product, Orders Pentagon," *Electronic Business,* Oct. 15, 1990, pp. 163–164.

23. Peter Burrows, "Commitment to Quality: Five Lessons You Can Learn from Award Entrants," *Electronic Business,* Oct. 15, 1990, pp. 56–58.

24. Gary S. Vasilash, "Hearing the Voice of the Customer," *Production,* Feb. 1989, pp. 66–68.

25. Ronald Fortuna, "Beyond Quality: Taking SPC Upstream," *Quality Progress,* June 1988, pp. 23–28. The author is manager of the Chicago office of Ernst & Young and observes that the control charts of SPC are considered one of the "Seven Old Tools" in Japan, along with Pareto analysis, cause-and-effect diagrams, data stratification, histograms, and scatter diagrams.

26. William Band and Richard Huot, "Quality & Functionality Equal Satisfaction," *Sales and Marketing Management in Canada (Canada),* March 1990, pp. 4–5.

27. For a more detailed and practical description of how to formulate and implement the method, see John R. Hauser and Don Clausing, "The House of Quality," *Harvard Business Review,* May–June 1988, pp. 63–73. Clausing, one of the authors, introduced QFD to Ford and its supplier companies in 1984 and the process was used in the successful design and introduction of the Taurus.

28. Dennis De Vera et al., "An Automotive Case Study," *Quality Progress,* June 1988, pp. 35–38.

29. At a Conference Board Quality Conference on April 2, 1990, A. F. Jacobson of 3M described how the Commercial Office Supply division brings customer expectations into the design process: "Let's say, they're going to develop an improved tape of some kind. They don't wait until the product is finished…or nearly finished…to take it to their customers. They take the idea to customers right at the beginning of the process. They ask customers what they want from a particular tape. Very often, they'll hear things like: 'Don't make it too sticky.' 'I want to be able to pull it off the roll easily.' and, 'It's no good unless I can write on it.' Now, collecting opinions is the easy part of the process. The tough part is converting these soft expectations into technical requirements. This is done on a matrix *before* the development process gets very far. These soft expectations are converted into technical specifications for, say, adhesion, roughness and reflectance."

30. Robert Haavind, "Hewlett-Packard Unravels the Mysteries of Quality," *Electronic Business,* Oct. 16, 1989, pp. 101–105.

31. Ronald M. Fortuna, "The Quality Imperative," *Executive Excellence,* March 1990, p. 1.

32. Steve Kaufman, "Quest for Quality," *Business Month,* May 1989, pp. 60–65.

33. Jack Byrd, Jr. and Mark D. Carter, "A Just-in-Time Implementation Strategy at Work," *Industrial Management,* March/April 1988, pp. 8–10. See also Ira P. Krespchin, "What Do You Mean by Just-in-Time?" *Modern Materials Handling,* Aug. 1986, pp. 93–95.

34. John H. Sheridan, "World-Class Manufacturing: Lessons from the Gurus," *Industry Week,* Aug. 6, 1990, pp. 35–41. Shingo also believes that JIT extends to plant maintenance, and with a participative environment operators will protect their own equipment.

35. There are a number of references that point out the benefits as well as the pitfalls of JIT. These sources also contain suggestions for implementation. See the following: Bruce D. Henderson, "The Logic of Kanban," *Journal of Business Strategy,* Winter 1986, pp. 6–12. The author describes how the technique can provide a competitive advantage. The need to rethink traditional practices is discussed in Lynne Perry, "Simplified Manufacturing Is Best," *Industrial Management,* July/Aug. 1986, pp. 29–30. The way that small manufacturers can adapt the technique is described in Byron Finch, "Japanese Management Techniques in Small Manufacturing Companies," *Production & Inventory Management,* Vol. 27 Issue 3, 3rd Quarter 1986, pp. 30–38. The need for continued quality control and other requirements is outlined in Mark R. Jamrog, "Just-in-Time Manufacturing: Just in Time for U.S. Manufacturers," *Price Waterhouse Review,* Vol. 32 Issue 1, 1988, pp. 17–29. The interface with other functions of the company is provided by R. Natarajan and Donald Weinrauch, "JIT and the Marketing Interface," *Production & Inventory Management Journal,* Vol. 31 Issue 3, 3rd Quarter 1990, pp. 42–46.

36. Joel E. Ross and Lawrence A. Klatt, "Quality: The Competitive Edge," *Management Decision (UK),* Vol. 24 Issue 5, 1986, pp. 12–16.

37. Peter Drucker, "The Emerging Theory of Manufacturing," *Harvard Business Review,* May/June 1990, p. 95.

38. James J. Webster, "Pulling—Not Pushing—For Higher Productivity," *Mechanical Engineering,* April 1988, pp. 42–44.

39. Gervase R. Bushe, "Cultural Contradictions of Statistical Process Control in American Manufacturing Corporations," *Journal of Management,* March 1988, pp. 19–31.

7

CUSTOMER FOCUS
and SATISFACTION

Quality begins and ends with the customer.

Joel Ross

Of all the Baldrige Award criteria, none is more important that customer focus and satisfaction. This category accounts for 300 of the 1000-point value of the award.

> This category examines the company's relationships with customers and its knowledge of customer requirements and of the key quality factors that drive marketplace competition. Also examined are the company's methods to determine customer satisfaction, current trends and levels of customer satisfaction and retention, and these results relative to the competition.[1]

The principles discussed in this chapter and in the entire book apply equally to both service and manufacturing firms. Judging from what is known about U.S. manufacturing and service firms, not many companies would not receive a grade of "A" for customer focus. A comprehensive study by the consulting firm Ernst & Young of 584 companies found that customer complaints were of "major or primary" importance in identifying new products and services among only 19 percent of banks and 26 percent of hospitals.

The widespread tendency to ignore complaints or track them and identify the cause(s) can have very serious consequences. This is particularly true in services, where it is estimated that for every complaint a business receives, there are 26 other customers who feel the same way but do not air their feelings to the company.[2]

Failure to identify the root cause of complaints means that reduction of variation in the causative process is more difficult. A customer unable to get through to a sales representative is evidence of a malfunction in the telephone procedure (process) or the sales and marketing function. Thus, it becomes necessary to tie the customer to the process.

Evidence indicates that part of the cause of this failure to close the customer-process loop is inadequate support from top management for the total quality management (TQM) infrastructure and a continued focus on the techniques of TQM, particularly statistical process control (SPC).

The Ernst & Young study mentioned previously found that quality-performance measures such as defect rates and customer satisfaction levels play a key role in determining pay for senior managers in only fewer than one in five companies. *Profitability* is still king. There is, of course, nothing wrong with a focus on cash flow and short-term profits, but long-term profit and market share require a base of satisfied customers that are retained by a focus on satisfaction. Some top executives may not like to believe the level or severity of customer complaints or may be offended by them. When Amtrak was criticized in the *Wall Street Journal* by a transportation analyst (Lind), the president of Amtrak responded (in the same paper):

▪ My own conclusion is that this [comment] is based on hopelessly incorrect assumptions about Amtrak and the railroad industry, and that Mr. Lind would be well advised to limit his comments and suggestions to the streetcar and transit business with which he is familiar and to avoid getting over his depth.[3]

While it may be true that the president of Amtrak is correct in this case, such an attitude expressed publicly could very well pervade the work force, who might perceive the message as justification for continuing the existing level of service.

Another reason for the lack of customer focus is the tendency of many firms to emphasize the techniques of TQM such as SPC and other outcome-oriented methods such as productivity and cost reduction. Again, these are desirable and necessary, but a singular emphasis on these areas is to put the cart before the horse.

The customer is not really interested in the sophistication of a company's process control, its training program, or its culture. The bottom line for the customer is whether he or she obtains the desired product. This truism is recognized by Deming, Juran, and Crosby.

PROCESS vs. CUSTOMER

Customer complaints are analogous to process variation. Both are undesirable and must be addressed. In both cases, the optimum output must be compared against an objective, a standard, or a benchmark. Both are integral parts of the quality improvement process. The integration of the customer and the process is shown conceptually in Figure 7-1.

From the company's point of view, customer satisfaction is the result of a three-part system: (1) company processes (operations), (2) company employees who deliver the product, and service that is consistent with (3) customer expectations. Thus, the effectiveness of the three-part system is a function of how well these three factors are integrated.

This concept is shown in Figure 7-2. The overlap (shaded area) represents the extent to which customer satisfaction is achieved. The

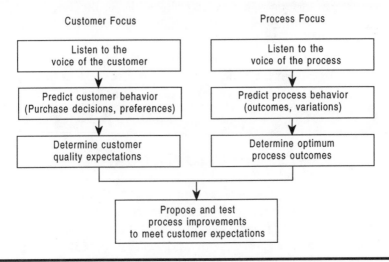

Figure 7-1 Integration of Customer and Process (Adapted from Dean E. Headley and Bob Choi, "Achieving Service Quality through Gap Analysis and a Basic Statistical Approach," *Journal of Services Marketing*, Winter 1992, p. 7.)

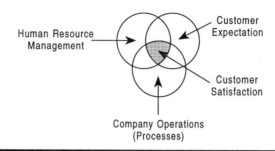

Figure 7-2   Customer Satisfaction: Three-Part System

objective is to make this area as large as possible and ultimately to make all three circles converge into an integrated system. The extent to which this condition is achieved depends on the effectiveness of (1) the process, (2) employees, and (3) determination of what constitutes "satisfaction." Like any system, control is necessary. Thus, standards are set, performance is measured, and variation, if any, is corrected.

> ■   Ritz-Carlton Hotel Company won the Baldrige Award in 1992. Many people thought that no hotel could do this because service in this industry is so difficult to measure and to deliver. The company meticulously gathers data on every aspect of the guest's stay to determine if the hotels are meeting customer expectations. Key to the research are the daily quality production reports that identify all problems and defects reported in each of 720 work areas. The data compiled range from the time it takes for housekeepers to clean a room to the number of guests who must wait in line to check in.[4]

INTERNAL CUSTOMER CONFLICT

Internal customers are also important in a TQM program. These are the people, the activities, and the functions within the company that are the customers of other people, activities, or functions. Hence, manufacturing is the customer of design, and several departments may be customers of data processing.

Conflict frequently arises between the needs of internal and external customers. In many cases, processes are designed to meet the needs of internal customers. Any "customer" who has been admitted to a hospital or outpatient service understands this. The registration process is designed to meet the needs of the admitting department, business office, or medical records. The result is a long wait to give information that will be provided again and again to personnel who represent admitting, the laboratory, finance, social work, and medicine. Who is the customer? Who is the beneficiary? Who is the recipient of the output? The patient gets the impression that he or she is a piece of raw material being moved along an amorphous assembly line known as health care.

It is not too difficult to identify other examples in both the private and public sectors. How about a university? It has been said that if you want to find out what kind of organization you are about to do business with, call on the phone!

A balance needs to be struck between the needs of these two customer groups. The solution is to determine the real needs of each and design the process to meet both.

DEFINING QUALITY

Supreme Court Justice Potter Steward once said that while he could not define "obscenity," he knew it when he saw it. Wrestling with a definition of quality is almost as difficult but necessary nevertheless. You cannot manage what you cannot measure.

The several dimensions of quality (performance, features, reliability, etc.) were discussed in a previous section.[5] However, the shortfall regarding product quality is that the services connected with it are so frequently overlooked. Good packaging, timely and accurate shipping, and the ability to meet deadlines matter as much as the quality of the product itself. Customers define quality in terms of their total experience with the company. Many companies approach customer satisfaction in a narrow way by confining quality considerations to the product alone.[6]

A QUALITY FOCUS

It is impossible to avoid the constant bombardment of "quality" and "satisfaction" messages in advertising on television and radio and in print media. Much of this advertising, and the actions to deliver the product or service, is little more than vague rhetoric. Even the popular phrases "satisfaction guaranteed" or "low price guaranteed" do not state what the customer is supposed to get for his or her purchase.

Some companies have attempted to improve this rhetoric by supplementing the message with additional definitions of satisfaction. McDonald's guarantees customer satisfaction with the pledge: "If you are not satisfied, we'll make it right or the next meal is on us." What does the phrase *make it right* mean? The question is whether this guarantee relates to product quality and customer satisfaction or is merely a promotion. Perhaps the slogan should be changed to "enjoyment guaranteed."

Many firms back up a satisfaction guarantee with promises of a reward if they fail to meet their own standards or those of the customer. Hampton Inns refunds your money. At Pizza Hut you get it free if not served in five minutes. Some firms give you a $5 bill. Delta Dental Plans of Massachusetts sends you a check for $50 if you get transferred from phone to phone while seeking an answer to an insurance question. Automobile dealers and manufacturers are fond of promoting "quality service" without defining just what this is. Some back it up with such specifics as towing service, free rides to work, or loaner cars when the customer's car is kept overnight.

There are two advantages to backing up a guarantee with some penalty for failure to deliver. It can cure employee apathy and bring quality to the attention of employees on a personal basis. It also may leave the buyer with a perception of dedication and thereby serve to retain what otherwise may have been a lost customer. These customers may say to themselves and others: "Well, my pizza was ten minutes late but they gave me a free one, so that proves they are serious about quality." Retaining this customer, who now has a better perception and higher expectation, may be worth the cost of the pizza and the foregone sale.

It should be remembered that any effort to tie the message of satisfaction to a failure-to-deliver penalty is ineffective if the variation or failure is not traced back to process improvement and the cause of the variation. Why was the pizza delivered late? Why was the customer shifted from phone to phone? Why did the dealer keep the car overnight? The variation-cause connection is identified by problem solving and the process improvement through process control.

Break Points

The need to improve customer satisfaction in *measurable* amounts is well known. But what is the measure and how much improvement is needed? If a customer is willing to stand in line for two minutes but finds five minutes unacceptable, anything between is merely satisfactory. Zero to one minute is outstanding. On-time delivery below 90 percent may be judged by customers as unacceptable, while over 98 percent is considered outstanding. Improvement programs should be geared toward reaching either a two-minute or five-minute range for standing in line and either 90 or 98 percent for delivery times. These are the market *break points,* where improving performance will change customer behavior, resulting in higher prices or sales volume. Forget the improvement program that targets one minute waiting in line or the delivery program that targets between 90 and 98 percent.

A Central Theme

Although individuals and teams may have targets that are directed at process improvement in their specific activities, a common theme or focus may integrate the many individual or group efforts that may have their own priority. At Motorola the theme is *six sigma;* at Hewlett Packard it is a *tenfold reduction in warranty expense.* At General Electric no part will be produced that cannot meet a *one-part-in-a-million* defect rate. At MBNA of America (a credit card company), the target is *customer retention.* In other companies it can be a reduction in defects or cycle time. Such a theme tends to be pervasive because so many individuals can relate their activities to it. It can serve to mobilize employees around an overall quality culture.

The DRIVER of CUSTOMER SATISFACTION

The benefits of having customers who are satisfied is well known and was outlined in Chapter 1. The issues in building customer satisfaction are to acquire satisfied customers, know when you have them, and keep them.[7]

The obvious way to determine what makes customers satisfied is simply to ask them. Before or concurrently with a customer survey, an audit of the company's TQM infrastructure needs to be made. IBM is

Table 7-1 Key Excellence Indicators for Customer Satisfaction

- ■ Service standards derived from customer requirements
- ■ Understanding customer requirements
  - ▪ Thoroughness/objectivity
  - ▪ Customer types
  - ▪ Product/service features
- ■ Front-line empowerment (resolution)
- ■ Strategic infrastructure support for front-line employees
- ■ Attention to hiring, training, attitude, morale for front-line employees
- ■ High levels of satisfaction—customer awards
- ■ Proactive customer service systems
- ■ Proactive management of relationships with customer
- ■ Use of all listening posts
  - ▪ Surveys
  - ▪ Product/service follow-ups
  - ▪ Complaints
  - ▪ Turnover of customers
  - ▪ Employees
- ■ Quality requirements of market segments
- ■ ▪ Surveys go beyond current customers
  - ▪ Commitment to customer (trust/confidence/making good on word)

one company that has identified the key excellence indicators for customer satisfaction. These key indicators are listed in Table 7-1.[8]

Despite the obvious need for customer input in determining new product/service offerings and improving existing ones, the widespread tendency is to determine perceived quality and perceived customer satisfaction based almost solely on in-house surveys.[9] Even when the company does attempt to get input, the survey may suffer from methodology shortcomings. Mailed questionnaires lose control over who responds, and respondents are less likely to reply if they are dissatisfied or if the name of the company or product is indicated. Just what is satisfaction? If the customer's expectation is low, satisfaction may be acceptable but perception will not improve. If perception is low but satisfaction is acceptable, how is this determined and what can be done? Suppose that 95 percent of respondents indicate satisfaction but do not perceive the product as one of the best. Survey results can be misconstrued and lead to complacency.[10]

▪ The hotel chain Ritz-Carlton, a Baldrige winner, relies on technology to keep comprehensive computerized guest history profiles on the likes and dislikes of more than 240,000 repeat guests. Researchers survey more than 25,000 guests each year to find ways in which the chain can improve delivery of its service.[11]

GETTING EMPLOYEE INPUT

Employee input can be solicited concurrent with customer research. It could help identify barriers and solutions to service and product problems, as well as serving as a customer-company interface.[12] Such surveys can help identify changes that may be necessary for quality improvement. In addition to customer-related considerations, employee surveys can measure (1) TQM effectiveness, (2) skills and behaviors that need improvement, (3) the effectiveness of the team problem-solving process, (4) the outcomes of training programs, and (5) needs of internal customers.[13]

▪ Corning Inc., a leader in the glassware industry, asked line and staff groups worldwide to assess themselves using the Baldrige criteria. Each group was to develop a few quality strategies that would address the most critical elements identified in the assessment. Measures, referred to a Key Result Indicators, that focused on evidence of customer deliverables and process outcomes were required.[14]

MEASUREMENT of CUSTOMER SATISFACTION

The accelerating interest in the measurement of customer satisfaction is reflected in the over 170 consulting firms that specialize in this activity.[15] Some firms use the "squeaky wheel" or "if it ain't broke, don't fix it" approach and measure customer satisfaction based on the level of complaints. This has a number of disadvantages. First, it focuses on the negative aspects by measuring dissatisfaction rather than satisfaction. Second, the measure is based on the complaints of a vocal few and may cause costly or unneeded changes in a process. As indicated at the beginning of this chapter, for every complaint, there are 26 others who feel the same way but do not air their feelings.

There are two basic steps in a measurement system: (1) develop key indicators that drive customer satisfaction and (2) collect data regarding the perceptions of quality received by customers.[16]

Key indicators of customer satisfaction are what the company has chosen to represent quality in its products and services and the way in which these are delivered. The building blocks that the system is designed to track are (1) expectations of the customer and (2) company perceptions of customer expectations.

In Chapter 1 a number of indicators for the physical product (e.g., reliability, aesthetics, adaptability, etc.) were identified. For service businesses or for services that accompany a product, the range of indicators depends on the nature of the service. One authority[17] has suggested that some important areas to consider are outcome, timeliness of the service, satisfaction, dependability, reputation of the provider, friendliness/courteousness of employees, safety/risk of the service, billing/invoicing procedures, responsiveness to requests, competence, appearance of the physical facilities, approachability of the service provider, location and access, respect for customer feelings/rights, willingness to listen to the customer, honesty, and an ability to communicate in clear language. These indicators, if appropriate and addressable, are converted to action items that reflect specific delivery systems where the product or service meets the customer. For example, in a bank customer needs and systems would combine to deliver short teller lines, friendly and courteous staff, ATMs that work, and low fees on accounts.

Data collection is required in order to identify the needs of customers and the related problems of process delivery. The data gathering process surveys both customers and employees. By including employees, customer needs and barriers to service can be identified, as well as recommendations for process improvement. Different orientations are emphasized for customers and employees; the former are asked for *their* expectations and the latter are asked what they think *customers* expect.

The ROLE of MARKETING and SALES

Marketing and sales are the functions charged with gathering customer input, but in many firms the people in these functions are unfamiliar with quality improvement.[18] Shortcomings in marketing as identified by critics include:[19]

- Partnering arrangements with dealers and distribution channels
- Focusing on the physical characteristics of products and overlooking the related services
- Losing a sense of customer price sensitivity
- Not measuring or certifying suppliers such as advertisers
- Failing to perform cost/benefit analyses on promotion costs
- Losing markets to generics and house brands

> ▪ According to one source, Motorola is a world-class producer of products but is less than world class in marketing. Historically, the company has been oriented toward engineering and technology. Its six sigma quality is well known. The publisher of *Technologic Computer Letter* says, "With many product lines Motorola has an extremely compelling story to tell but it is used to hiding its light under a bushel and does not make its advantages heard."[20]

Quality and customer satisfaction have not played an important role in the sales function (*process*). Consider the stereotype of a salesperson. He or she is detail (rather than process) oriented and trained in technical product knowledge (rather than customer knowledge). Salespeople are feature oriented: "We've got six models, four colors, and it comes with a money-back guarantee." They are trained and rewarded for getting new customers, as opposed to retaining existing ones.

The SALES PROCESS

According to Hiroshi Osada of the Union of Japanese Scientists and Engineers (JUSE), TQM needs to begin with the salespeople.[21] Yet TQM has migrated to the sales force in only a few companies.[22] Even fewer perceive sales as a *process* that lends itself to analysis and improvement for customer satisfaction. To repeat a previous caveat, "If you can't measure it, you can't manage it." Another can be added: "You can't measure it if its not a process." Both of these cliches are as true for sales and marketing as for any other process. The objective is quality outcomes. In order for TQM to become a part of sales and marketing, managers and employees must move toward a deeper understanding of its processes—selling, advertising, promoting, innovating, distribution, pricing, and packaging—*as they relate to customer satisfaction.*

Marketing applications need not be confined to the marketing department. Other functions can borrow these techniques to improve the satisfaction of external and internal customers. A brokerage firm should care not only about sending accurate statements on time, but should also be concerned with whether the statement format fits the customer's needs. In an issue of *Marketing News,* Research Professor Eugene H. Fram of the Rochester Institute of Technology suggests the following types of non-traditional marketing extensions:[23]

- **Adapted** marketing refers to a non-marketing function that adapts traditional techniques. Relationship selling is an example. If a human resource department sends the same recruiter each year to a campus, this person can use principles of relationship selling to further company goals. This type of selling can also be used within the organization and between departments. The classic conflict between production (cost) and sales (delivery) can be reduced.

- **Morale** marketing can improve morale. Consider what the terms "Team Taurus" or "Team Xerox" did for morale in those companies.

- **Sensitivity** marketing borrows from the basic marketing principle which says that one must understand the customer's needs in order to fulfill them and to build long-term relationships. In a marketing sense, individuals, groups, and departments are better able to achieve quality and productivity if they are sensitive to the needs, concerns, and priorities of both internal and external customers.

SERVICE QUALITY and CUSTOMER RETENTION

Customer defection is a problem and customer retention an opportunity in both manufacturing and service firms. Manufacturers have generally been good about measuring satisfaction with products, but now they are moving into service areas. The publicity surrounding the Baldrige Award accounts for much of this. Other reasons relate to the size and growth of service industries and the growing importance of service as a means of strategically competing in the marketplace.

Service industries are playing an increasingly important role in a nation's economy. Over 75 percent of the working population in the United States is employed in the service sector and the percentage is growing. When this employment is combined with service jobs in the manufacturing sector, it becomes evident that the importance of services is increasing. Many executives feel that the management of services is one of the most important problems they face today. Yet most of us know from personal experience that the quality of services is declining, despite the efforts of some companies to improve it.

Because so many services are intangible, the interaction between employees and customers is critical. Chase Manhattan Bank realizes that an employee's ability to meet or exceed customer expectations when conducting a routine transaction influences the customer's satisfaction with the organization. In fact, this interaction influences satisfaction more than the actual product or service obtained. The one-on-one or face-to-face contact between the customer and the deliverer of the service (nurse, flight attendant, retail clerk, restaurant server) is extremely important.

Manufacturers are careful to measure material yield, waste scrap, rework, returns, and other costs of poor quality processes. Service companies also have these costs, which are reflected in the cost of customers who will not come back because of poor service. These are customer defections and they have a substantial impact on cost and profits. Indeed, it is estimated that customer defections can have a greater impact than economies of scale, market share, or unit cost.[24] Despite this, many companies fail to *measure* defections, determine the *cause* of defections, and improve the *process* to reduce defections.

CUSTOMER RETENTION and PROFITABILITY

What is the ultimate desired outcome of customer focus and satisfaction? Is it achieving profit in the private sector or productivity in the public or non-profit sectors? The answer must be yes. Oddly enough, however, an accurate cause-and-effect relationship has yet to be established between profit and customer satisfaction. This is due, in part, to the difficulty of measuring satisfaction and relating it to profit. However, there is a proven relationship between *customer retention and profit*.

One way to put a value on customer retention is to assign or estimate a "lifetime retention value," the additional sales that would result if the customer were retained. Taco Bell calculates the lifetime value of a retained customer as $11,000. An automobile dealer believes that the lifetime value of retaining a customer is $300,000 in sales.[25] Conversely, MBNA America (a credit card company) has found that a 5 percent improvement in customer *defections* increases its average customer value by 125 percent.

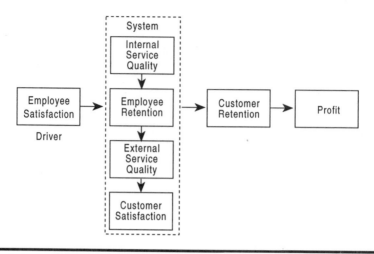

Figure 7-3 Profitability and Customer Retention

The system for improving *customer retention* and profit is illustrated in Figure 7-3. The drivers are *employee satisfaction* and employee retention. The system components are

▪ **Internal service quality,** which establishes and reinforces a climate and organization culture directed toward quality.

▪ **Employee retention,** which is achieved through good human resource management practices and organization development methods such as teams, job development, and empowerment. Employee retention depends on employee satisfaction, which in turn can be related to external service and customer satisfaction.

▪ **External service quality,** which is delivered through the organization's quality infrastructure.

▪ **Customer satisfaction** and follow up, in order to reduce customer defections and improve retention.

To reiterate, there is a proven relationship between customer retention and profit.

BUYER–SUPPLIER RELATIONSHIPS

Almost every company purchases products, supplies, or services in an amount that frequently equals around 50 percent of its sales.

Traditionally many of these companies have followed the "lowest bidder" practice where price is the critical criterion. The focus on price, even for commodity products, is changing as companies realize that careful concentration of purchases, together with long-term supplier–buyer relationships, will reduce costs and improve profits.[26] Deming realized this and suggested that a long-term relationship between purchaser and supplier is necessary for best economy.[27] If a buyer has to rework, repair, inspect, or otherwise expend time and cost on a supplier's product, the buyer is involved in a "value/quality-added" operation, which is not the purpose of having a reliable supplier. In that never-neverland of the perfect buyer–supplier relationship, no rework or inspection is necessary.

A partnership arrangement is emerging between a growing number of manufacturers and suppliers. At Eastman Kodak, the Quality Leadership Process (QLP) has improved the company's production processes, reduced overall manufacturing costs, and improved quality by transforming traditional manufacturer–supplier roles. Because one-half of all components used in manufacturing are supplied by outside vendors, realignment of the supplier base has become a central strategy of QLP.[28]

> ▪ Motorola has advanced the supplier–customer relationship further than most companies. The system is based on a basic economic principle: whenever someone buys from someone else, there is a mutually beneficial transaction and pleasing both sides is important. With this in mind, Motorola has begun to market itself as a customer. The company's director of materials and purchasing says, "If the sauce is good for the goose, it should be good for the gander, and we are genuinely trying to cooperate, collaborate and do some strategic things with our suppliers. Our goal is to become a world-class customer and that means that it is important for use to learn what the buyer needs to do in order for suppliers to see us as a world-class customer."[29]

Several guidelines will help both the supplier and customer benefit from a long-term partnering relationship:

▪ **Implementation of TQM by both supplier and customer.** Many customers (e.g., Motorola, Ford, Xerox) are requiring suppliers to operationalize the basic principles of TQM. Some have even required the supplier to apply for the Baldrige Award. This joint effort provides a common language and builds confidence between both parties.

- **Long-term commitment to TQM and to the partnering relationship between the parties.** This may mean a "life cycle" relationship that carries partnering through the life cycle of the product, from market research and design through production and service.

- **Reduction in the supplier base.** One or more automobile companies have reduced the number of suppliers from thousands to hundreds. Why have ten suppliers for a part when the top two will do a better job and avoid problems?

- **Get suppliers involved in the early stages of research, development, and design.** Such involvement generates additional ideas for cost and quality improvement and prevents problems at a later stage of the product life cycle.

- **Benchmarking.** Both customer and supplier can seek out and agree on the best-in-class products and processes.

How does one become a quality supplier? This of course depends on the criteria of the buyer, but it is reasonably safe to assume that if the following criteria are met, a company can reasonably expect to be classified in the quality category. The following criteria are required to be certified as *quality* in the automobile industry:

1. *Management philosophy* of the CEO should support TQM
2. Techniques of *quality control* should be in place (SPC, etc.)
3. Desire for a long-term *life cycle relationship*
4. Best-in-class *inventory and purchasing systems*
5. *Facilities* should be up to TQM standards
6. *Automation* level should meet quality standards
7. *R & D and design* should support customer expectations
8. Willing to *share costs*

QUESTIONS for DISCUSSION

7-1 Describe how a program directed toward customer focus and satisfaction interacts with

- The information and analysis component of the TQM approach
- Strategic quality planning
- Human resource development and management
- Management of process quality

7-2 Select a function or activity (e.g., design, order processing, accounting, data processing, engineering, market research) and identify a measure of quality that you would expect if you were an *internal* customer of that function or activity.

7-3 Choose a specific product or service in a particular industry and devise an action plan for obtaining customer input and feedback. How would the information generated by such a plan be used for process improvement?

7-4 Illustrate how a firm might focus on *internal* product or service specifications rather than customer expectations and desires.

7-5 Choose a product or service and list four or five characteristics that you as a customer would want and expect. Based on your experience, do you think that the firm will deliver?

7-6 How would you establish a system to measure customer satisfaction?

ENDNOTES

1. *Malcolm Baldrige National Quality Award Criteria—1993,* Washington, D.C.: U.S. Department of Commerce, National Institute of Standards and Technology, 1993, p. 29.
2. "Satisfaction-Action," *Marketing News,* Feb. 4, 1991, p. 4.
3. "Management: Quality Programs Show Shoddy Results," *Wall Street Journal,* May 14, 1992, Section B, p. 1.

4. Edward Watkins, "How Ritz-Carlton Won the Baldrige Award," *Lodging Hospitality,* Nov. 1992, p. 23.

5. See David A. Garvin, *Managing Quality,* New York: The Free Press, 1988, pp. 49–59. Garvin has defined the eight dimensions of quality as performance, features, reliability, conformance, durability, serviceability, aesthetics, and perceived quality. Computer-maker NCR goes to great expense to define quality as appropriateness, reliability, aesthetics, and usability. Industrial designers have given the company a silver medal for design. For a comprehensive report on how product design enhances profits and market share, see a special report, "Hot Products: How Good Design Pays Off" (cover story), *Business Week,* June 7, 1993, pp. 54–78.

6. Oren Harari, "Quality Is a Good Bit-Box," *Management Review,* Dec. 1992, p. 8.

7. Gerald O. Cavallo and Joel Perelmuth, "Building Customer Satisfaction, Strategically," *Bottomline,* Jan. 1989, p. 29.

8. These indicators are taken from class material in an IBM in-house workshop, "MDQ (Market Driven Quality) Workshop." The company was kind enough to share this class material with the author and several other professors from the College of Business, Florida Atlantic University. For this, I thank them.

9. It was found, for example, that in the hospital industry fewer than 5 percent of referring physicians play a prominent role in identifying new service opportunities. Nearly 40 percent of U.S. hospitals indicate that senior management "always or almost always" takes the dominant role in identifying new services." U.S. hospitals seek minimal input from patients. *The international Quality Study, Healthcare Industry Report,* American Quality Foundation and Ernst & Young, 1992.

10. For some ideas on getting customer input, see Joel E. Ross and David Georgoff, "A Survey of Productivity and Quality Issues in Manufacturing: The State of the Industry," *Industrial Management,* Jan./Feb. 1991.

11. Edward Watkins, "How Ritz-Carlton Won the Baldrige Award," *Lodging Hospitality,* Nov. 1992, p. 24.

12. Luane Kohnke, "Designing a Customer Satisfaction Measurement Program," *Bank Marketing,* July 1990, p. 29.

13. Kate Ludeman, "Using Employee Surveys to Revitalize TQM," *Training,* Dec. 1992, pp. 51–57.

14. David Luther, "Advanced TQM: Measurements, Missteps, and Progress through Key Result Indicators at Corning," *National Productivity Review,* Winter 1992/ 1993, pp. 23–36.

15. Lynn G. Coleman, "Learning What Customers Like," *Marketing News,* March 2, 1992, pp. CSM-1–CSM-11. This article contains a directory of 170 customer satisfaction measurement firms.

16. For a more detailed description of how to establish a measurement system, see J. Joseph Cronin, Jr. and Steven A. Taylor, "Measuring Service Quality: A Reexamination and Extension," *Journal of Marketing,* July 1992, pp. 55–68. See also Luane Kohnke, "Designing a Customer Measurement Program," *Bank Marketing,* July 1990, pp. 28–30; Gerald O. Cavallo and Joel Perelmuth, "Building Customer Satisfaction, Strategically," *Bottomline,* Jan. 1989, pp. 29–33.

17. Dean E. Headley and Bob Choi, "Achieving Service Quality through Gap Analysis

and a Basic Statistical Approach," *Journal of Services Marketing,* Winter 1992, pp. 5–14. This is a good primer on gap analysis and the use of basic statistical techniques.

18. Joe M. Inguanzo, "Taking a Serious Look at Patient Expectations," *Hospitals,* Sep. 5, 1992, p. 68. This article points out that there is very little employee involvement in measuring satisfaction and practically none from patients.

19. Allan J. Magrath, "Marching to a Different Drummer," *Across the Board,* June 1992, pp. 53–54.

20. B. G. Yovovich, "Becoming a World-Class Customer," *Business Marketing,* Sep. 1991, p. 16.

21. Dick Schaaf, "Selling Quality," *Training,* June 1992, pp. 53–59.

22. John Franco, president of Learning International in Stamford, Connecticut, has conducted a series of round table discussions with sales executives. He reported: "When we ask participants how many of them are from companies that have a quality movement underway, we find about half do. But when we ask them whether that effort has migrated to the sales force, fewer than 10 percent say it has." Dick Schaaf, "Selling Quality," *Training,* June 1992, pp. 53–59.

23. Eugene H. Fram and Martin L. Presberg, "TQM Is a Catalyst for New Marketing Applications," *Marketing News,* Nov. 9, 1992.

24. Frederick F. Reicheld and W. Earl Sasser, Jr., "Zero Defections: Quality Comes to Services," *Harvard Business Review,* Sep./Oct. 1990, pp. 105–111. Reprint No. 90508.

25. Harvard Business School video series, "Achieving Breakthrough Service," Boston: Harvard Business School, 1992.

26. Robert D. Buzzell and Bradley Gale, *The PIMS Principles,* New York: The Free Press, 1987, p. 62. The data from over 3000 strategic business units show that concentrating purchases *improves* profitability, at least up to a point. "The positive net effect of a moderate degree of purchase concentration suggests that the efficiency gains that can be achieved via this approach to procurement are usually big enough to offset the disadvantages that might be expected as a results of an inferior bargaining position."

27. W. Edwards Deming, *Out of the Crisis,* Cambridge, Mass.: Massachusetts Institute of Technology, Center for Advanced Engineering Study, 1986, p. 35.

28. Joseph P. Aleo, Jr., "Redefining the Manufacturer–Supplier Relationship," *Journal of Business Strategy,* Sep./Oct. 1992, pp. 10–14.

29. B. G. Yovovich, "Becoming a World-Class Customer," *Business Marketing,* Sep. 1991, p. 29.

8

BENCHMARKING

Benchmarking is a way to go backstage and watch another company's performance from the wings, where all the stage tricks and hurried realignments are visible.

Wall Street Journal

In Joseph Juran's 1964 book *Managerial Breakthrough,* he asked the question: "What is it that organizations do that gets results so much better than ours?" The answer to this question opens the door to *benchmarking,* an approach that is accelerating among U.S. firms that have adopted the total quality management (TQM) philosophy.

The essence of benchmarking is the continuous process of comparing a company's strategy, products, and processes with those of world leaders and best-in-class organizations in order to learn how they achieved excellence, and then setting out to match and even surpass it. For may companies, benchmarking has become a key component of their TQM programs. The justification lies partly in the question: "Why re-invent the wheel if I can learn from someone who has already done it?" C. Jackson Grayson, Jr., chairman of the Houston-based American Productivity and Quality Center, which offers training in benchmarking and consulting services, reports an incredible amount of interest in benchmarking. Some of that interest may be explained by the criteria for the Malcolm Baldrige Award, which includes "competitive comparisons and benchmarks."[1]

The EVOLUTION of BENCHMARKING

The method may have evolved in the 1950s, when W. Edwards Deming taught the Japanese the idea of quality control. Other American management innovations followed. However, the method was rarely used in the United States until the early 1980s, when IBM, Motorola, and Xerox became the pioneers. The latter company became the best-known example of the use of benchmarking.

Xerox

The company invented the photocopier in 1959 and maintained a virtual monopoly for many year thereafter. Like "Coke" or "Kleenex," "Xerox" became a generic name for all photocopiers. By 1981, however, the company's market share shrank to 35 percent as IBM and Kodak developed high-end machines and Canon, Ricoh, and Savin dominated the low-end segment of the market. The Xerox vice president of copier manufacturing remarked: "We were horrified to find that the Japanese were selling their machines at what it cost us to make ours...we had been benchmarking against ourselves. We weren't looking outside." The company was suffering from the "not invented here" syndrome, as Xerox managers did not want to admit that they were not the best.

The company instituted the benchmarking process, but it met with resistance at first. People did not believe that someone else could do it better. When faced with the facts, reaction went from denial to dismay to frustration and finally to action. Once the process began, the company benchmarked virtually every function and task for productivity, cost, and quality. Comparisons were made for companies both in and outside the industry. For example, the distribution function was compared to L. L. Bean, the Freeport, Maine catalog seller of outdoor equipment and clothing and everyone's model of distribution effectiveness.

By the company's own admission, it would probably not be in the copier business today if it were not for benchmarking. Results were dramatic:

- Suppliers were reduced from 5000 to 300.
- "Concurrent engineering" was practiced. Each product development group has input from design, manufacturing, and service from the initial stages of the project.
- Commonality of parts increased from about 20 percent to 60 to 70 percent.

■ Hierarchical organization structure was reduced, and the use of cross-functional "Teams Xerox" was established.

■ Results included:
  • Quality problems cut by two-thirds
  • Manufacturing costs cut in half
  • Development time cut by two-thirds
  • Direct labor cut by 50 percent and corporate staff cut by 35 percent while increasing volume

It should be noted that all of these improvements were not the direct result of benchmarking. What happened at Xerox (and what happens at most companies) is that in adopting the process, the climate for change and continuous improvement followed as a natural result. In other words, benchmarking can be a very good intervention technique for positive change.

Ford

The entire automobile industry may have undergone substantial change as a result of Ford's Taurus and Sable model cars. Operating performance and reliability were significantly improved, and the gains were recognized by U.S. car buyers as well as others in the industry. "Team Taurus," a cross-functional group of employees, was empowered to bring the car to market and was given considerable authority to act outside of the normal company bureaucracy.

The team defined 400 different areas that were considered important to the success of a mid-size car. A best-in-class competition was chosen for each area. Fifty different mid-sized car models were chosen. Few were Ford models. Based on the 400 benchmarks, specific teams were assigned responsibility to meet or beat the best-in-class for each area of performance, and 300 features were "copied" and incorporated into the car design. Target dates were set for beating the remaining features. "Quality Is Job One" became the fight song for Ford employees.

The Taurus was, and is, a resounding success. Some auto analysts credit the Taurus experience with the partial resurgence of quality in the U.S. automobile industry. The benchmarking process provided additional benefits. During the examination of competitors' features, valuable insights into the design process were gained. Cycle time was reduced. Buyer–supplier relationship was improved as supplier input was solicited for the design. All manufacturing processes were improved as a by-product of the benchmarking process.

Motorola

In the early 1980s, the company set a goal of improving a set of basic quality attributes *tenfold* in five years. Based on *internal* benchmarking, the goal was reached in three years. The company then began to look outside, sending teams to visit competitor plants in Japan. To their chagrin, the teams found that Motorola would have to improve its tenfold improvement level another two to three times just to match the competition.

Borrowing process benchmarks from companies as diverse as Wal-Mart, Benetton, and Domino's Pizza, the company now routinely fields benchmarking requests from those same Japanese companies it toured the first time around.[2]

The ESSENCE of BENCHMARKING

The process is more than a means of gathering data on how well a company performs against others both in and outside the industry. It is a method of identifying new ideas and new ways of improving processes and hence meeting customer expectations. Cycle time reduction and cost cutting are but two process improvements that can result. The traditional approach of measuring defect rates is not enough. The ultimate objective is *process improvement* that meets the attributes of customer expectation. This improvement, of course, should meet both strategic and operational needs.

A properly designed and implemented benchmarking program will take a total system approach by examining the company's role in the supply chain, looking upstream at the suppliers and downstream at distribution channels. How competitive are suppliers in the world market and how well are they integrated into the company's own core business processes—product design, demand forecasting, product planning, and order fulfillment.[3]

BENCHMARKING and the BOTTOM LINE

There are two basic points of view regarding how to get started in benchmarking. One minority view maintains that an *initial* action plan that tries to match the techniques used by world-class performance may actually make things worse by doing too much too soon. A three-year study of 580 global companies conducted by the management consulting firm Ernst & Young concluded that it may be best to start measuring existing financial performance measures. Two key measures are return on assets (which is simply after tax income divided by total assets) and value added per employee. Value added is sales minus the costs of materials, supplies, and work done by outside contractors. Labor and administrative costs are not subtracted from sales to arrive at value added.[4]

The focus on financial results is not recommended by the majority of executives familiar with the benefits of benchmarking. Some believe that it is easy to be fooled by financial indicators that lull the company into thinking that it is doing well when what in reality occurs is a transitory financial phenomenon that may not hold up over the longer term. A more important payoff is quality processes that lead to a quality product.

Robert C. Camp headed up the now-famous study at Xerox in which the buzzword "benchmarking" was coined in 1980. When asked whether the best work practices necessarily improve the bottom line, he replied: "The full definition of benchmarking is finding and implementing best practices in our business, practices that meet customer requirements. So the flywheel on finding the very best is, 'Does this meet customer requirements?' There is a cost of quality that exceeds customer requirements. The basic objective is satisfying the customer, so that is the limiter."[5]

The BENEFITS of BENCHMARKING

Given the considerable effort and expense required for effective benchmarking, why would an organization embark on such an effort? The answer is justified by three sets of benefits.

Cultural Change

Benchmarking allows organizations to set realistic, rigorous new performance targets, and this process helps convince people of the credibility of these targets. This tends to overcome the "not invented here" syndrome and the "we're different" justification for the status quo. The emphasis on looking to other companies for ideas and solutions is antithetical to the traditional U.S. business culture of individualism.

Robert Camp, the former Xerox guru quoted earlier, indicates that the most difficult part for a company that is starting the process is getting people to understand that there may be people out there who do things better than they do. According to Camp, overcoming that myopia is extremely important.

Performance Improvement

Benchmarking allows the organization to define specific gaps in performance and to select the processes to improve. It provides a vehicle whereby products and services are redesigned to achieve outcomes that meet or exceed customer expectations. The gaps in performance that are discovered can provide objectives and action plans for improvement at all levels of the organization and promote improved performance for individual and group participants.

Human Resources

Benchmarking provides a basis for training. Employees begin to see the gap between what they are doing and what best-in-class are doing. Closing the gap points out the need for personnel to be involved in techniques of problem solving and process improvement. Moreover, the synergy between organization activities is improved through cross-functional cooperation.

STRATEGIC BENCHMARKING

It is paradoxical that two AT&T divisions (AT&T Network Systems Group, Transmission Systems Business Unit, and AT&T Universal Card Services) were 1992 winners of the Baldrige Award. Like several other winners, the company has turned this win into an advantage and organized a separate operation to market this expertise. Training is the product offered by the AT&T Benchmarking Group of Warren, New Jersey.[6] The process is illustrated in Figure 8-1.

The paradox is that ten years earlier, in 1983, AT&T was convinced that it could be a major player in the computer industry. The company owned Bell Laboratories, the largest R & D facility in the world, and had extensive experience in the manufacture of telecommunications equipment, a related product.

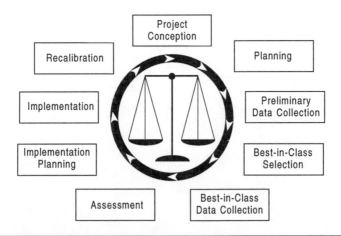

Figure 8-1 AT&T Benchmarking Process

Five years after entering the industry and after losing billions of dollars, the company was still trying to be a significant player in the market. The near disaster could be traced directly to the company's failure to (a) realize that the key success factors in the industry included sales, distribution, and service (functions that AT&T had very little experience in) and (b) conduct *strategic benchmarking* against such best-in-class competitors as IBM and Compaq. Moreover, the company apparently failed to define its market segment, the criteria used for customer purchasing decisions, and how the company's product could be differentiated in the chosen segment. If, for example, IBM, Compaq, or AT&T wanted to benchmark NCR, they would find that NCR has gone to great expense to define the criteria of product quality as "usability, aesthetics, reliability, functionality, innovation and appropriateness."[7]

One way to determine how well you are prepared to compete in a segment and to help define a best-in-class competitor is to construct a key success factor (KSF) matrix similar to the one shown in Figure 8-2. Following this determination, a matrix such as the hypothetical one shown in Figure 8-3 can be constructed to measure market differentiation criteria against competitors. Note that the criteria for comparison are based on the customer's purchase decision. This type of strategic analysis can be followed by one involving specific processes—operational benchmarking. Strategy drives performance and hence quality. Indeed, quality can and should become the central theme of strategy.

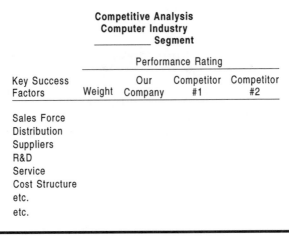

Figure 8-2 Key Success Factor Matrix

Note that Figures 8-2 and 8-3 can be used to benchmark best-in-class *outside* the industry.[8]

OPERATIONAL BENCHMARKING

This category focuses on a particular activity within a company's functional operations and then identifies ways to emulate or improve on the practices of best-in-class. Whereas strategic benchmarking is

Computer Industry
_____ **Segment**
Customer's Purchase Decision

		Performance Rating			
Criteria	Weight	Our Company	Competitor #1	Competitor #2	etc.
Reliability					
Performance					
Features					
Durability					
Service					
Software					
etc.					
etc.					

Figure 8-3 Measuring Market Differentiation Criteria against Competitors

largely concerned with the macro analysis of the environment, the industry, and the competitors, operational benchmarking is more detailed in terms of data gathering and the rigor of the analysis. Most of the focus is on cost and differentiation. Because the customer's purchasing decision (PD) is a function of price and differentiation, it is necessary to differentiate through *quality* [PD = $f(P \times Q)$] and improve price through *cost* reduction. Both lead to an analysis of the cost and activity chains of interconnected processes.

The scope of benchmarking extends to both strategic and operational processes. The scope of these two categories of benchmarking at Westinghouse (a Baldrige winner) is displayed in Table 8-1.

Table 8-1   How Westinghouse Uses Competitive Benchmarking Data

Process benchmarks	Product benchmarks
Categories	**Categories**
Assessment	Development
Performance	Features
Technology	Functionality
Financial	Architecture
Organizational	Availability
Development	**Marketing**
Goals	Target markets
Analysis	Market positioning
Countermeasures	Price strategies
Implementation	
Improvement	**Sales**
Gap analysis	Product positioning
Targets	Bid responses
Countermeasures	Customer talks
Comparisons and competitive analysis	**Comparisons and competitive analysis**
Scope	Features
Complexity	Functionality
Technology	Architecture
Performance	Availability
Cost	Market position
Strength/weakness	Price
Documentation	Strength/weakness
	Documentation

The BENCHMARKING PROCESS

There is no standard or commonly accepted approach to the benchmarking process. Each consulting group[9] and each company[10] uses its own method. Whatever method is used, the major steps involve (1) measuring the performance of best-in-class relative to critical performance variables, (2) determining how the levels of performance are achieved, and (3) using the information to develop and implement a plan for improvement. These steps are discussed in further detail in the following sections.

Determine the Functions/Processes to Benchmark

Those functions or processes that will benefit the most should be targeted for benchmarking. It is wise to choose those that absorb the highest percent of cost and contribute the greatest role in differentiation, always thinking in terms of process improvements that will have a positive impact on the customer's purchasing decision. Because no company can excel at everything, it is necessary to delineate targets. Benchmarking "manufacturing," for example, is much too broad and the subject is too ill-defined. If the elements to be benchmarked cannot be framed, data gathering is not focused and subsequent actions may be destructive.

Many companies focus their efforts on product comparisons. In manufacturing industries this may mean product tear-downs (e.g., Ford, Xerox) and re-engineering of design standards and assembly processes. This approach should take second place to improving time to market, first-time quality of design, and design for purchasing effectiveness, which are the primary drivers of both quality and cost. Of course, these actions should be undertaken after customer satisfaction has been defined with customer input.

▪ The health care industry provides an example of the potential for cost and quality improvement. For one procedure alone, coronary artery bypass grafts (CABGs, DRGs 106-7), Americans paid for more than 130,000 in 1991. Of the patients treated, 6,033 died. Ancillary charges alone reached $2.67 billion. Baxter Healthcare Corporation of Deerfield, Illinois, which benchmarked CABGs in ten hospitals, calculated that $1.57 billion in ancillary charges alone could be saved if all hospitals benchmarked the processes of the benchmarked ten.[11]

Select Key Performance Variables

Functions, activities, and processes can be measured in terms of specific output measures of operations and performance. In general, these measures fall into four broad categories.

Cost and productivity, such as overhead costs and labor efficiency. Total dollars per unit or per ton is a starting point in manufacturing. Other variables might include production yield of raw material, direct labor per unit produced, etc. Unless the project team begins with total costs before it breaks them down by process or activity, some very important overhead charges may be neglected when benchmarked against firms with different accounting systems. See Chapter 10 (Productivity and Quality) for additional measures.

Comparing one company's financial statements and cost breakdowns against those of another would be a good method for a "me-too" strategy *if* access were available to the detailed statements of a competitor or the best-in-class and *if* they were based on similar accounting methodology. These are two big "if's." A better way is to identify the underlying cost *drivers* of the many functions and activities that, when combined, make up total costs. For example, raw material costs may be driven by sales, purchase volume, source, or freight; direct labor by wage and benefit rates, skilled vs. unskilled, or union vs. non-union; indirect labor by the ratio of direct to indirect, salary levels, and so on.

■ A team at Mercy Hospital in San Diego decided to benchmark medical records because the activity represented the largest portion of clinical support. The team left a benchmarking visit to a sister hospital empty-handed because they found that the two hospitals were quite different in this activity. A team member commented: "They weren't equivalent to us at all. It didn't do the functions we did, it wasn't open 24 hours a day like us, and it was more decentralized—a lot of what we do, they do in various other departments and clinics"

Timeliness. Often overlooked, timeliness is a major factor in internal processes as well as customer satisfaction. The measure is frequently expressed in cycle time or turnaround time such as time to fill an order or time to answer the phone. Some manufacturing executives have been known to visit automobile races to measure pit stops as benchmarks for set-up time or line changeover time.

Differentiation and quality. Measures of differentiation and quality are needed for both processes and product. Quality measures should capture the errors, defects, and waste attributable to an entire process and express them relative to the total output achieved. Defects tend to cascade down a chain of processes, becoming increasingly expensive to correct.

Differentiation and quality of product are essentially the same, because quality is what differentiates a product. The variables should include any factors that affects a customer's purchasing decision (see, for example, Figure 8-3).

Business processes. These are the processes not directly related to product design, production, sales, and service. They include the many staff and internal service activities that are costed under general and administrative (G&A) expense. One has only to look at the organizational chart to identify areas for cost reduction and for improvement of productivity and quality. Human resources, data processing, accounts receivable, marketing services, maintenance, security, data center, warehousing, public relations...the list goes on. Many companies have had severe cash flow and profit problems due to a failure to control the cost and output of these business or support processes. Whereas direct labor and material costs may make up the largest segment of total costs in a manufacturing firm and can be benchmarked more easily, G&A costs are more elusive and more difficult to measure; however, they represent fertile ground for improvement. Another area is internal quality and internal customers. A good place to start may be to use the techniques of activity analysis and activity-based costing.

IDENTIFY the BEST-in-CLASS

This is a major step in the benchmark analysis. The objective is to identify companies whose operations are superior, the so-called best-in-class, so that the company's own operations can be targeted.

The quickest way to identify excellent performers is simply to visit some companies that have won the Baldrige Award. A lot could be learned in a hurry, but these companies may not have the time or may not have similar processes. Other sources include (1) available databases, (2), sharing agreements between companies, and (3) out-of-industry companies.

Databases are an expanding source of comparison information. The most current and most comprehensive of these is maintained by the Houston-based American Productivity & Quality Center (AP&QC). Some of the chief difficulties that organizations encounter are identifying top-performing companies in specific functions and finding companies that have already conducted studies in specific areas. Helping others overcome these difficulties is the role of the AP&QC. It serves as a central networking source and has the support of top benchmarkers.

The cost of membership in the AP&QC ranges from $6,000 to $12,500, depending on the number of employees. Dissemination of benchmarking information is through face-to-face networking meetings, electronic bulletin boards, and on-line access to abstracts of company benchmark studies.[12]

As the popularity of benchmarking accelerates, so does the number of consortium efforts among industry peer groups. For example, a number of hospitals have formed the MECON-PEER database to provide information and analysis software for examining individual operations and compare them with similar operations nationwide. Some of the participants have discovered an additional use for the database: putting muscle into a budget squeeze and justifying additional resources based on benchmarking activities of peers.

Even universities are emerging as benchmarkers. Oregon State University pioneered the process in the academic world, and their success led to the creation of NACUBO, a database of the National Association of College and University Business Officers.

A number of companies are also developing *in-house* databases. This is particularly effective in large multi-division companies, where economies of scale in data sharing can be achieved. One such company is AT&T. The extent of the competitive benchmarking data maintained by the Network Systems Group for use by all company divisions is shown in Table 8-1 (see earlier).[13]

Cooperative sharing agreements between companies is another source of best-in-class identification. Members of the agreement may or may not be competitors and may or may not be in the same industry. DEC, Xerox, Motorola, and Boeing joined forces to standardize benchmarking procedures in training.

Out-of-industry companies may be the best source of information for many firms in the early or intermediate stages of project implementation. A benchmark planner at Johnson & Johnson suggests that 90 percent of all opportunities for breakthrough improvement lie in studying practices outside the industry. Perishable food companies often teach other manufacturers about supply-demand balancing, demand forecasting, production scheduling, and distribution management. Pharmaceutical companies are quite knowledgeable in production record keeping, quality assurance, and batch traceability.

Although many companies are mistakenly paranoid about sharing strategic and operating information, many others are not. Most Baldrige winners and applicants and people from many best-in-class companies are just regular people and are proud of what they have accomplished.

> ■ When Mid-Columbia Medical Center of The Dalles, Oregon got serious about TQM and benchmarking, they formed the "MCMC University." The director, dubbed "Professor," decided to benchmark the training function and spent five days taking notes at Disney University, and then went on to attend Ritz-Carlton's training session for a week. "They were flattered," the "Professor" said. "We were the only people who had ever asked them if we could attend." Training videos were supplied by Northwest Tool & Die Company, Disney, Harley-Davidson, and Johnson Sausage.

Table 8-2 contains a selected list of companies noted for their best practices in the functions shown.

MEASURE YOUR OWN PERFORMANCE

At this step in the process, your own performance should have been pre-measured; otherwise, there is nothing to compare against the benchmarking data. Moreover, data analysis of best-in-class may proceed aimlessly unless the benchmarker understands what information is being sought.

Having determined with some degree of accuracy the performance of the target firm and the extent of your own performance, it follows that an analysis of the *gap* between the two is necessary. The trickiest part of the process is to compare internal and external data on an equivalent basis. This does not mean that both sets of data must be comparable in the same exact form.

Table 8-2   Selected Best-Practice Companies

Company	Function
American Airlines	Information systems (long line)
American Express (Travel Services)	Billing
AMP	Supplier management
Benetton	Advertising
Disney World	Optimum customer experience
Domino's Pizza	Cycle time (order and delivery)
Dow Chemical	Safety
Emerson Electric	Asset management
Federal Express	Delivery time
General Electric	Management processes
GTE	Fleet management
Herman Miller	Compensation and benefits
Hewlett-Packard	Order fulfillment
Honda	New product development
IBP	Productivity
L. L. Bean	Distribution
3M	Technology transfer
Marion Merrell Dow	Sales management
Marriott	Admissions
MBNA America	Customer retention
Merck	Employee training
Milliken	Cross-functional processes
Motorola	Flexible manufacturing
NEXT	Manufacturing excellence
Ritz-Carlton	Training
Travelers	Health care management
US Sprint	Customer relations
Wal-Mart	Information systems
Xerox	Benchmarking

Performing a "gap analysis" of the variation with the benchmarked process involves the problem-solving process treated in Chapter 6. This analysis will reveal:

- The extent, the size, and the frequency of the gap
- Causes of the gap; why it exists

■ Available methods for closing the gap and reaching the performance level of the benchmarked process

ACTIONS to CLOSE the GAP

Once the *cause(s)* of the gap is determined through problem analysis, alternative courses of action to close the gap become evident. Selecting the right alternative course of action is a matter of rational decision making. Among the criteria for weighing the courses of action are time, cost, technical specifications, and, of course, quality. It should be added here that the best source of information on closing the gap may be the best-in-class, because that company has already experienced what the benchmarking organization is going through.

The action plan lists each action step, the time of completion, the person responsible, and the cost, if appropriate. The results expected from each action step should also be listed in order to provide a measure of whether the objective or output of each step is achieved.

The action plan itself represents a process and lends itself to the basics of process control. Hence monitoring, feedback, and recalibration are required.

PITFALLS of BENCHMARKING

Curt W. Reimann, who heads the Baldrige Award program at the National Institute of Standards and Technology, finds that a lot of people think benchmarking is "instant pudding." It will not improve performance if the proper infrastructure of a total quality program is not in place. Indeed, there is significant evidence that it can be harmful. Unless a corporate culture of quality and the basic components of TQM (such as information systems, process control, and human resource programs) are in place, trying to imitate the best-in-class may very well disrupt operations.

Other potential pitfalls include the failure to:

■ Involve the employees who will ultimately use the information and improve the process. Participation can lead to enthusiasm.

■ Relate process improvement to strategy and competitive positioning. Design to factors that affect the customer's purchasing decision.

▪ Define your own process before gathering data or you will be overwhelmed and will not have the data to compare your own process.

▪ Perceive benchmarking as an ongoing process. It is not a one-time project with a finite start and complete date.

▪ Expand the scope of the companies studied. Confining the benchmarking firms to your own area, industry, or to competitors is probably too narrow an approach in identifying excellent performers that are appropriate for your processes.

▪ Perceive benchmarking as a means to process improvement, rather than an end in itself.

▪ Set goals for closing the gap between what is (existing performance) and what can be (benchmark).

▪ Empower employees to achieve improvements that they identify and for which they solve problems and develop action plans.

▪ Maintain momentum by avoiding the temptation to put study results and action plans on the back burner. Credibility is achieved by quick and enthusiastic action.

QUESTIONS for DISCUSSION

8-1 What benefits can be gained from benchmarking?

8-2 Identify two or three functions or activities, other than product characteristics, that could be benchmarked by

▪ A manufacturer

▪ A service company

8-3 How can benchmarking become an intervention technique for organizational change?

8-4 Summarize some actions taken by Xerox, Ford, and Motorola while implementing their benchmarking programs.

8-5 What are the pros and cons of benchmarking based on financial performance?

8-6 Select an industry and list three or four key success factors (e.g., advertising, distribution, engineering, sales) for that industry. Which firm(s), in your opinion, would be appropriate to benchmark?

ENDNOTES

1. Rick Whiting, "Benchmarking: Lessons from the Best-in-Class," *Electronic Business,* Oct. 7, 1991, pp. 128–134. This article provides a good justification for benchmarking and the principles behind it.

2. Bob Gift and Doug Mosel, "Benchmarking: Tales From the Front," *Healthcare Forum,* Jan./Feb. 1993, pp. 37–51.

3. A. Steven Walleck, "Manager's Journal: A Backstage View of World-Class Performers," *Wall Street Journal,* Aug. 26, 1991, Section A, p. 10. This article contains good examples of benchmarking applications in several companies.

4. See "Quality," a special report in *Business Week,* Nov. 30, 1992, p. 66. This report suggests various benchmarking measures for three types of firms: the novice, the journeyman, and the master.

5. Adrienne Linsenmeyer, "Fad or Fundamental?" *Financial World,* Sep. 17, 1991, p. 34.

6. The address of the group is 10 Independence Blvd., Warren, NJ 07059. Florida Power & Light Company, the only U.S. winner of the Japanese Deming Prize, formed Qualtec, a consulting group offering services in quality management.

7. *Wall Street Journal,* May 26, 1992, Section C, p. 15.

8. Perhaps the largest *strategic* database is the PIMS (Profit Impact of Marketing Strategy) collection maintained at the Strategic Planning Institute in Cambridge, Massachusetts. The database contains the strategic and financial results of over 3000 strategic business units. A member firm can search for strategic "look-alike" firms and benchmark the determinants of good or not so good performance. See Robert D. Buzzell and Bradley T. Gale, *The PIMS Principles,* New York: The Free Press, 1987. See also Bradley T. Gale and Robert D. Buzzell, "Market Perceived Quality: Key Strategic Concept," *Planning Review,* March/April 1989, pp. 6–48.

9. For example, Kaiser Associates, Inc. has a seven-step process which is outlined in a company publication, *Beating the Competition: A Practical Guide to Benchmarking,* Vienna, Va.: Kaiser Associates, 1988.

10. For example, Alcoa's steps include (1) deciding what to benchmark, (2) planning the benchmarking project, (3) understanding your own performance, (4) studying others, (5) learning from the data, and (6) using the findings. See Alexandra Biesada, "Benchmarking," *Financial World,* Sep. 17, 1991, p. 31.

11. Bob Gift and Doug Mosel, "Benchmarking: Tales from the Front," *Healthcare Forum,* Jan./Feb. 1993, p. 38.

12. David Altany, "Benchmarkers Unite," *Industry Week,* Feb. 3, 1992, p. 25.

13. Taken from a company brochure entitled *A Summary of AT&T Transmission Systems: Malcolm Baldrige National Quality Award Application.* AT&T's database contains data from over 100 companies and over 250 benchmarking activities for key processes such as hardware and software development, manufacturing, financial planning and budgeting, international billing, and service delivery. Over 20,000 entries describe benchmarking trips or visits with internal and external customers. Sources of competitive benchmarking information include customers, visits to other companies, trade shows and journals, professional societies, standards committees, product brochures, outside consultants, and installation data.

9

ORGANIZING for TOTAL QUALITY MANAGEMENT

*If you still believe in hierarchy, job descriptions and func-
tional boundaries, and are not experimenting with new
approaches to boundaryless/networked/virtual organizations
engaged in ever-changing partners, you are already in deep
yogurt.*

Tom Peters
Forbes

Synthesizing quality values and policies into every person's job and
every operation is a complex task that must be supported by an
appropriate organizational infrastructure. Management texts universally
define organizing as a variation of a statement such as "the process of
creating a structure for the organization that will enable its people to
work together effectively toward its objectives."[1] Thus, the process
recognizes a structural as well as a behavioral or "people" dimension.

This chapter is concerned with the macro dimension of organization:
the overall approach the company might take to establish a quality
infrastructure. The micro dimension (organizing the "quality department" or the duties of the "top quality manager") is technical in nature
and beyond the scope of this book. Both Deming[2] and Crosby[3] treat this
in some detail.

Historically, organizations have tended to focus on the classical principles of specialization of labor, delegation of authority, span of control (a limited number of subordinates), and unity of command (no one works for two bosses). The result in many cases was the traditional pyramidal organization chart, cast in stone and accompanied by budgets, rules, procedures, and the chain of command hierarchy. Task specialization was extreme in some cases. The classic bureaucracy thus emerged.

Prior to the current emergence of total quality management (TQM) in the early 1980s, responsibility for quality was vague and confusing. Executive management grew detached from the idea of managing to achieve quality. The general work force had no stake in increasing the quality of its products and services. Quality had become the business of specialists—product specification engineers and process control statisticians who determined acceptable levels of product variability and performed quality control inspection on the factory floor.

Today, it is generally recognized that there are two prerequisites for a TQM organization. The first is a quality attitude that pervades the entire organization. Quality is not just a special activity supervised by a high-ranking quality director.[4] This attitude (culture, vision) was examined in Chapter 2 and is largely a challenge for top management. The second prerequisite is an organizational infrastructure to support the pervasive attitude. Companies must have the means and the structure to set goals, assign them to appropriate people, and convert them to action plans. People must be aware of the importance of quality and trained to accomplish the necessary tasks.

ORGANIZING for TQM: The SYSTEMS APPROACH

A system can be defined as an entity composed of interdependent components that are integrated for achievement of an objective. The organization is a social system comprised of a number of components such as marketing, production, finance, research, and so on. These organizational components are activities that may or may not be integrated, and they do not necessarily have objectives or operate toward achievement of an objective. Thus, synergism, a necessary attribute of a well-organized system, may be lacking as each activity takes a parochial view or operates independently of the others. This lack of synergism cannot continue under the TQM approach to strategic

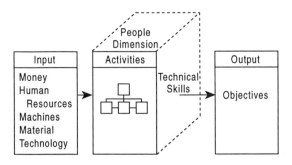

Figure 9-1 The Organization System

management because interdependency across functions and departments is a necessary precondition.

The concept of an organizational system is shown in Figure 9-1. Inputs to the system are converted by organization activities into an output. Indeed, the sole reason for the existence of the organization and each activity within it is to add value to inputs and produce an output with greater value. A measure of this conversion of inputs into outputs is known as productivity, and the ratio of output to input must be a positive number if the system is to survive in the long run.

The activities of the organization are subsystems of the whole, but are also individual systems with inputs and outputs that provide input to other systems such as customers and other internal activities. This *chain* of input/output operations is depicted in Figure 9-2.

Despite the simplicity of the concept, it most often fails in practice. Activity supervisors and individuals within activities do not understand the objective or results of their "subsystem," nor can they define their output in measurable terms. When asked to define the output of their jobs, they will answer: "I am responsible for maintenance," or "I work in finance," or "my job is to ship the product." In each case these are statements of activity and not output, objective, or results expected. *Quality* output is stated in such vague terms as "do a good job" or "keep the customer happy." People can describe what they do (activity) but not what they are supposed to get done (objective or result). They may be very efficient at doing things right but ineffective in doing the right things. This failure is critical to organization output as well as structure.

Michael Porter, in his excellent book *Competitive Advantage,*[5] has taken the systems theory a major practical step forward with his concept

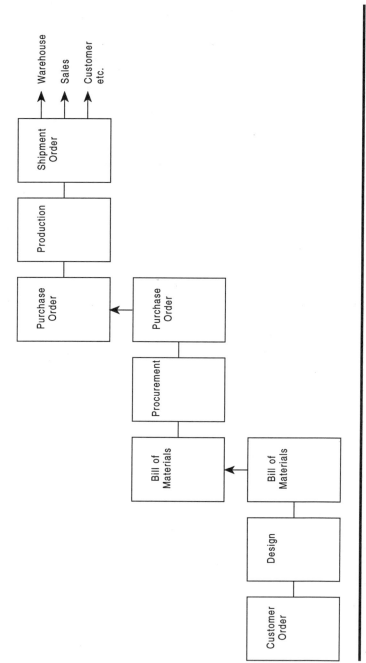

Figure 9-2 Chain of Subsystems

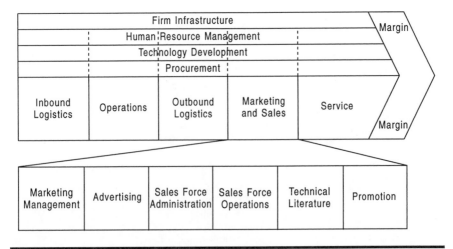

Figure 9-3 Subdividing a Generic Value Chain

of *the value chain*. He suggests that "competitive advantage (in this case quality) cannot be understood by looking at a firm as a whole. It stems from the many discrete activities a firm performs in designing, producing, marketing, delivering, and supporting its product." While Porter's concept is expanded to include any of the many sources of competitive advantage, the value chain concept will be used here to focus on the organizational structure for TQM.

The discrete activities of an organization can be represented using the generic value chain shown in Figure 9-3. Note that the activities or organizational functions are comprised of primary and support activities, which may or may not be changed from those listed in Figure 9-3 depending on the firm's industry and its particular strategy. Selected examples of chain activities from Porter's book are summarized in Table 9-1.

Customers, channels, and suppliers also have value chains, and the firm's output of product or service becomes an input to the customer's value chain. The firm's differentiation and its competitive advantage depend on how the activities in its value chain relate to the needs of the customer, channel, or supplier. If quality has been chosen as a competitive advantage, it now remains to determine the customer's value chain and how the product or service can add value to the customer's system. Following this determination, the value chain should be organized into the required discrete activities, each one of which

Table 9-1 Chain Activities

Primary activities	Support activities
Inbound logistics Materials handling Warehousing Inventory control Vehicle scheduling Returns to suppliers	**Procurement** Dispersion of the procurement function throughout the firm
	Technology development Efforts to improve products and processes
Operations Machining	**Human resource management** Recruiting
Management Packaging Assembly Maintenance Testing	Hiring Training Development Compensation
Outbound logistics Material handling Order processing Scheduling Finished goods warehouse	**Firm infrastructure** Supports entire chain General management Planning Finance, accounting Quality management
Marketing and sales Advertising Promotion Sales force Pricing	
Service Installation Repair Training Parts supply	

can improve the quality of the output for the purpose of meeting the customer's expectations. Before asking what you can do for the customer, ask what the customer expects to accomplish. The answer forms the basis for a quality organization. In this regard, it should be kept in mind that there are linkages between a firm's value chain and those of its customers, as well as downstream linkages with channels and suppliers. An excellent example of this is Wal-Mart, where a key competitive advantage was achieved through the value chain activity of technology development; in Wal-Mart's case, it was the sophisticated computer-based information system that improved the output of many other activities such as distribution, purchasing, and warehousing.

■ A spokesperson for Winnebago Industries, manufacturer of motor homes, concludes, "You must pick the right distribution network. In our case, it is our dealers. We believe we are only as strong as our dealer network. They are our first, last, primary, and most critical link to our end customers."[6]

■ Globe Metallurgical of Cleveland, the first small company to win the Baldrige Award, realized the importance of suppliers in their own value chain. Globe's management determined that the most effective method of assuring compliance with statistical process control and quality approaches in the suppliers' facilities would be to visit each supplier location with a quality improvement team and to train the hourly employees at each location. The program is a vital aspect of Globe's quality system.[7]

■ BSQ Group, an architectural firm in Tulsa, Oklahoma, designs and constructs stores for Wal-Mart. Although the firm's immediate customer is Wal-Mart, they organize their value chain to go downstream with linkages to Wal-Mart's customer: "Many people believe that quality is generally in the eyes of the beholder. Well, in the case of Wal-Mart, that beholder is the store's customer. They are the ones that are helping us define the quality standards that we currently strive to present and it's with them in mind that we begin our study."[8]

ORGANIZING for QUALITY IMPLEMENTATION

The traditional approach to organization sees the process as a mechanical assemblage of functions and activities without a great deal of attention to strategy and desired results. The process takes the product as given and groups the necessary skills and activities into homogeneous functions and departments. This approach to building an organization structure has been criticized by Peter Drucker: "What we need to know are not all the activities that might conceivably have to be housed in the organization structure. What we need to know are the load-bearing parts of the structure, the *key activities*."[9]

Key activities will differ depending on the nature of the organization, its products, and its strategy. What is a key activity in one may not be in another. Advertising may be a key activity in the value chain of Coca-Cola, but not in Boeing Aircraft, where design is the key activity. Back office activity may be a key activity in Merrill Lynch, but not in McDonald's. Firms frequently fail to prioritize or identify key activities in the value chain because of a tendency to organize around the chart of accounts. Some firms focus on those activities where cost, rather than quality or other source of differentiation, is the major consideration.

The value chain concept provides a systematic way to identify the key activities necessary for quality differentiation and a way to group them into homogeneous departments and functions. Indeed, an organization structure that corresponds to the value chain is the most economic and effective way to deliver quality and therefore achieve a competitive advantage.

It should be noted that the Quality Assurance Department is generally not the load-bearing key activity when organizing for TQM. Quality assurance activities can be found in nearly every function of the company if these functions are viewed as links in the value chain. Any activity or function is a potential source of quality differentiation. The ill-defined or elusive word "quality" may be too narrow if it focuses on product or service alone. Moreover, such limited focus may exclude the many other activities that impact the customer's value chain. Not only those functions normally classified as "line" but a variety of "staff" functions as well can be the source of quality in the organization structure. Consider the following sample activities:

Activity	Value to customer
Purchasing	Improved cost and quality of product
Engineering and design characteristics	Unique product
Manufacturing	Product reliability
Order processing	Response time
Service	Customer installation
Scheduling	Response time
Inspection	Defect-free product
Spare parts	Maintenance
Human resources	Customer training

By listing the activities of the organization and comparing them to a value chain such as Figure 9-3, one can see the many potential ways that quality differentiation can be achieved. It should also be noted that these activities can lower customer costs as well.

Production of quality does not stop when the product leaves the factory. Distribution and service are part of the production process. Careful identification of customer value will reveal a number of other opportunities for quality differentiation. For example, buyers and potential customers frequently perceive value in ways they do not understand or because of incomplete knowledge. Scanning a daily newspaper or magazine quickly reveals the many way that both manufacturers and service firms signal subjective, qualitative measures of quality. Do you buy Pepsi Cola for taste or brand image? Do you contemplate the purchase of a Volvo for performance or long life and safety? Consulting and accounting firms signal quality by the appearance and presumed professionalism of employees. Banks are known to build impressive facilities to indicate quality. Charles Revson, formerly of Revlon, once said, "I'm not selling cosmetics, I'm selling hope." The several criteria that the buyer may use to make a buying decision means that there may be an equal number of activities that become *key* activities in the creation of customer value. Porter provides several illustrative signaling criteria,[10] to which firm examples and organization activities that become key in delivery of the criteria have been added here:

Criteria	Firm example	Activity involved
Reputation	Appliances	Advertising
Appearance	Apparel	Design
Label	Athletic shoes	Graphics
Facilities	Bank	Maintenance
Time in business	Whiskey	Distribution
Customer list	Magazine publisher	Marketing
Visibility of top management	Consumer products	Hot line

Of course, having signaled a particular criterion to buyers and potential buyers, it is necessary to deliver as promised, measure the effectiveness of the criterion, and keep customer feedback communication lines open to ensure satisfaction.

Delivery of quality products or services depends on how well the many activities of the company are organized and integrated. The measurement of effectiveness is fundamental to the TQM process (see Chapter 7). It now remains to organize for customer feedback, another *key activity* that impacts other functions and activities throughout the organization.

Measuring customer satisfaction, or dissatisfaction, is an essential but often overlooked activity. What happens when a customer chooses a bank's trust department based on the criterion of experienced personnel, only to be shunted off to a recent college graduate or ignored by a "customer representative"? Research indicates that customers who are satisfied with a bank's quality will tell, on the average, three other people, while those who are dissatisfied will tell eight or nine others about poor quality.[11] How does a customer feel when returning an item under warranty only to be patronized by a retail clerk. One survey found that for every problem incident reported to corporate headquarters, there are at least 19 other similar incidents which simply were not reported or which were handled by the retailer or the front line without being recorded. Most companies spend 95 percent of their resources handling complaints and less than 5 percent analyzing them.

There is a strong correlation between consumer satisfaction with response to problems or questions and the likelihood of purchasing another product from the same company.[12] Yet few customers bother to complain, and of those who do, only a small fraction reach top management. What is needed is the institutionalization of customer service throughout the organization as a key activity to be performed by everyone. Despite this evident need, many companies have neither the activities nor the supporting policies. For many who do, there is a conflict between organization and policies that may have an opposite effect. Covertly measuring quality by using mystery shoppers, holding motivational meetings which employees perceive as paternalistic and patronizing, and paying for sales rather than service are among those policies that may conflict with the need to provide quality products and service.[13] It may be difficult for employees to be quality conscious in the face of policies that discourage this attitude.

The PEOPLE DIMENSION: MAKING the TRANSITION from a TRADITIONAL to a TQM ORGANIZATION

The typical company (Figure 9-4a) operates with a vertical, functional organizational structure based on reporting relationships, budgeting procedures, and specific and detailed job classifications.[14] Departmentation is by function, and communication, rewards, and

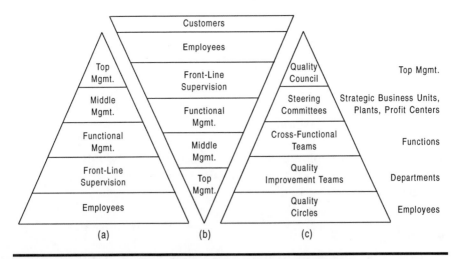

Figure 9-4 Transition from Traditional to TQM Organization

loyalties are functionally oriented. Processes are forced to flow vertically from the top down, creating costly barriers to process flow.

The systems approach to organizing suggests three significant changes, one conceptual and two requiring organizational realignment:

■ The concept of the inverted organizational chart

■ A system of intra-company internal quality

■ Horizontal and vertical integration of functions and activities

The Inverted Organizational Chart

If you've seen one organizational chart, you've seen them all: the symmetrical pyramid with the chairman at the top and the cascading of authority to successive levels (fourteen at General Motors) until the functions are shown near the bottom of the chart. Front-line supervisors are rarely shown and non-supervisory personnel almost never appear.

Where are the front-line supervisor and the employees? These are the people who deliver quality to the customer. In the eyes of the customer, they *are* the company. The sports fan cares not for the owner or the manager. The players deliver the quality. And so it is with the flight attendant, the bank teller, the auto mechanic, the sales person explaining a product, the person answering the telephone...even the college professor.

Perhaps it is time to put first things first. To make the transition from traditional to TQM management, it may be desirable to *conceptualize* a new organizational chart. Invert the existing one (Figure 9-4b) and put the customer at the top, followed by the employees and front-line supervisors. These are the deliverers of quality. This concept does not change the hierarchy and flow of authority, but the boss is no longer the boss in the old-fashioned sense. He or she is now a facilitator, a coach, and an integrator, whose job is to remove barriers that prevent subordinates from doing their jobs. The same role now falls on middle and top management. Quality is now the responsibility of everyone and not just the quality assurance department.

Internal Quality

The Juran Institute of Wilton, Connecticut delivers a program called "Managing Business Process Quality," which is a technique for executing cross-functional quality improvement among intra-company functions and activities.[15] A key factor in this approach is an organization-wide focus on the customer, including both *internal* and *external* customers. An enlarged definition of quality should be used to embrace all business processes, rather than just manufacturing.

The systems approach, by definition, requires the integration of organizational activities for achievement of a common goal. This goal, under the TQM form of organization, remains the satisfaction of customer requirements, but customers are now considered to be both outside as well as within the organization.[16] The process applies whether relating to a final customer or an internal customer; it is a participative process involving supplier and customer in an active dialogue. Examples include:

■ Metropolitan Life Insurance Company has made a major commitment to improve quality by implementing a *horizontal management* approach that is built on management commitment, employee involvement, and knowledge of internal suppliers.[17]

■ Campbell USA has aimed its latest quality emphasis, its "Quality Proud" program, at the administrative and marketing activities of the company. Job descriptions, promotions, pay, and bonuses for all employees are linked to the results of the new program.[18]

As a major step in its transformation to a total quality organization, DEC asked each of its 125,000 employees to answer in writing the following questions:

1. What business process are you involved in?
2. Who are your customers (that is, the next step in the processes you are involved in)?
3. Who are your suppliers (that is, the preceding step in the processes you are involved in)?
4. Are you meeting the expectations of your customers?
5. Are your suppliers meeting your expectations?
6. How can the processes be simplified and waste eliminated?[19]

▪ DEC reported that this simple survey had a massive impact. In the short run, countless redundant activities were discovered and eliminated. In the long run, DEC employees now think in terms of meeting both internal and external customer expectations. (This concept is also illustrated in Figure 9-2.)

Aside from the obvious benefits of improvements in quality, productivity, and cost, a system of internal customer quality is important for a number of other reasons:

▪ External customer satisfaction cannot increase unless internal customer satisfaction does.
▪ No quality improvement effort can succeed without employee buy-in and proactive participation.
▪ Focus on internal quality promotes a quality and entrepreneurial culture.
▪ An understanding of internal quality policy is an aid in communication and decision making.
▪ It is a significant criterion in the Malcolm Baldrige National Quality Award (Section 5.6).

ROLES in ORGANIZATIONAL TRANSITION to TQM

Members of a successful organization need a sound understanding of their roles during the transition to a TQM program. People at all levels require orientation as to how they will be impacted under the new philosophy of employee involvement. The improvement process involves a group of complementary activities that provide an environment conducive to improvement of performance for both employees and managers. Each level has a role to play.

The role of **top management** is critical. Many of the most successful companies launched their programs by creating a quality council or steering committee (Figure 4-4c) whose members comprise the top management team. Some multi-division companies encourage a council in each division or strategic business unit (SBU). The council provides a good vehicle for management to demonstrate its leadership in the quality initiative. At Motorola the CEO, who is also the Chief Quality Officer of the corporation, chairs the Operating and Policy Committee in all-day meetings twice each quarter.[20]

Opinions differ as to who should lead or coordinate the TQM effort. One source suggests a new role similar to that of a financial controller, a role that is justified on the basis that quality is now a strategic business planning and management function.[21] Others disagree and suggest that the company should avoid setting up a quality bureaucracy headed by a high-profile quality director. There is general agreement that it should not be headed by a staff department such as personnel or quality assurance. The process should be line led and given back to the business managers who implement it on a daily basis. To reiterate, quality should not be led by a non-line manager.

The major changes are strategic and organizational and have been outlined in this and previous chapters. It now remains for top management to manage the transition.[22]

The role of **middle managers** has traditionally been an integrative one. They are the drivers of quality and the information funnel for change both vertically and horizontally—the go-between for top management and front-line employees. They implement the strategy devised by top management by linking unit goals to strategic objectives. They develop personnel, make continuous improvement possible, and accept responsibility for performance deficiencies.[23]

Front-line supervision has been called the missing link in TQM.[24] At Federal Express, a Baldrige winner, the communication effort is focused on the front-line supervisors because most employees report directly to them. The company realizes that the real purveyors of quality are the employees, and a basic quality concept is candid, open, two-way communication.

Supervisors can make or break a quality improvement effort. They are called upon to provide support to employee involvement teams and create a climate that builds high levels of commitment in groups and individuals.

Quality assurance and the quality professional are faced with good news and bad news as TQM emerges as the load-bearing concern of company strategy. On the one hand, the accelerating emphasis on quality has given them more visibility, and in some cases the reporting relationships have moved to higher levels in the organization. On the other hand, they may now be perceived as a staff support function as quality becomes more widespread and led by line managers.

Philip Crosby indicates that the quality professional must become more knowledgeable about the process of management.[25] The limited tools of inspection techniques and statistical process control have become less important as the more sophisticated approaches of TQM begin to pervade all functions and activities, rather than just manufacturing.

SMALL GROUPS and EMPLOYEE INVOLVEMENT

In a *Harvard Business Review* article, David Gumpert described a small "microbrewery" where the head of the company attributed their success to a loyal, small, and involved work force. He found that keeping the operation small strengthened employee cohesiveness and gave them a feeling of responsibility and pride.[26]

This anecdote tells a lot about small groups (hereafter called teams) and how they can impact motivation, productivity, and quality. If quality is the objective, employee involvement in small groups and teams will greatly facilitate the result because of two reasons: motivation and productivity.

The theory of motivation, but not necessarily its practice, is fairly mature, and there is substantial proof that it can work. By oversimplifying a complex theory, it can be shown why team membership is an effective motivational device that can lead to improved quality.[27]

Teams improve productivity as a result of greater motivation (Table 9-2) and reduced overlap and lack of communication in a functionally based classical structure characterized by territorial battles and parochial outlooks. There is always the danger that functional specialists, if left to their own devices, may pursue their own interests with little regard for the overall company mission. Team membership, particularly a cross-functional team, reduces many of these barriers and encourages

Table 9-2   Team Membership and Motivation

Motivating factors	Team membership
Job development (the work)	
Vertical loading	Provides responsibility
Job closure	Team members see results
Feedback	Self-established goals
Achievement	Targets set by teams
Growth/self-development	Training, more responsibility
Recognition	By peers and supervisors
Communication	Team is vehicle for communication (see Chapter 3)

an integrative systems approach to achievement of common objectives, those that are common to both the company and the team. There are many success stories. To cite a few:

■ Globe Metallurgical, Inc., the first small company to win the Baldrige Award, had a 380 percent increase in productivity which was attributed primarily to self-managed work teams.[28]

■ The partnering concept requires a new corporate culture of participative management and teamwork throughout the entire organization. Ford increased productivity 28 percent by using the team concept with the same workers and equipment.[29]

■ Harleysville Insurance Company's Discovery program provides synergism resulting from the team approach. The program produced a cost saving of $3.5 million, along with enthusiasm and involvement among employees.[30]

■ At Decision Data Computer Corporation middle management is trained to support "Pride Team."[31]

■ Martin Marietta Electronics and Missiles Group has achieved success with performance measurement teams (PMTs).[32]

■ Publishers Press has achieved significant productivity improvements and attitude change from the company's process improvement teams (PITs).[33]

■ Florida Power & Light Company, the utility that was the first recipient of the Deming Prize, has long had quality improvement teams as a fundamental component of their quality improvement program.[34]

TEAMS for TQM

The several subsystems or components of a TQM approach were examined in previous chapters. The most critical of these components is employee involvement, and it is the one around which the management system of TQM should be based. It is the most important of the components of TQM and also the most complex. Consider the analogy of an iceberg. Approximately 10 percent of an iceberg is visible, while 90 percent is hidden from view. Imagine that the organizational chart is an iceberg. The visible 10 percent is top management and functional management. The 90 percent, where the true potential for quality exists, is comprised of front-line supervision and non-management employees. Does it not make good sense to tap into the 90 percent which represents a reservoir of ideas for quality and productivity improvements? The vehicle for doing this is some form of *team*.

A 1989 General Accounting Office study found that over 80 percent of all companies had implemented some form of employee involvement.[35] However, the statistic is misleading because responding companies considered a suggestion system as an employee involvement program, which is hardly a systems approach or a linking vehicle. Moreover, the methods most likely to have enduring effects are those that covered the least percentage of employees.

Quality Circles

The most widespread form of an employee involvement team is the quality circle, defined as "a small group of employees doing similar or related work who meet regularly to identify, analyze, and solve product-quality and production problems and to improve general operations."[36] Although the concept has had some success in white-collar operations, the major impact has been among "direct labor" employees in manufacturing, where concerns are primarily with quality, cost, specifications, productivity, and schedules. By their very nature, quality circles were limited to concerns of the small group of members and few cross-functional problems were considered.

The major growth of the circles occurred in the late 1970s and early

1980s, as thousands of companies adopted the concept. Like so many previous movements (e.g., management by objectives, value analysis, zero-based budgeting), however, the concept never met expectations and widespread abandonment resulted. As many as 50 percent of Fortune 500 companies disbanded their circles in the 1980s.[37] The major reason for failure was a general lack of commitment to the concept of participation and the lack of interest and participation by management.[38] From a TQM perspective, quality circles lack the prerequisites of integration with strategy, company goals, and management systems. Organizations can go beyond using circles by creating task forces, work teams, and cross-functional teams.[39]

Task teams are a modification of the quality circle. The major differences are that task teams can exist at any level and the goal or topic for discussion is given, whereas in quality circles members are generally free to choose the problems they will solve. Task teams with the best chance for success are those that represent an extension of a pre-existing, successful quality circle program.[40]

Self-managing work teams are an extension of quality circles but differ in one major respect: members are empowered to exercise control over their jobs and optimize the efficiency and effectiveness of the total process rather than the individual steps within it. Team members perform all the necessary tasks to complete an entire job, setting up work schedules and making assignments to individual team members. Peer evaluation is another characteristic.[41]

Cross-Functional Teams

▪ Computer manufacturer DEC has integrated a range of proven TQM techniques into its program, including cross-functional process improvement teams. One element is strictly home-grown. DELTA (DEC Employees Leveraged Team Activities) is a sophisticated, closed-loop suggestion system designed to discover and address problems. Under DELTA, only an employee who makes a suggestion can dispose of it. He or she also has the responsibility of working with other employees to implement or reshape the suggestion in order to determine whether it is feasible. Thus, DELTA empowers employees and promotes team building, two essential elements of quality management.[42]

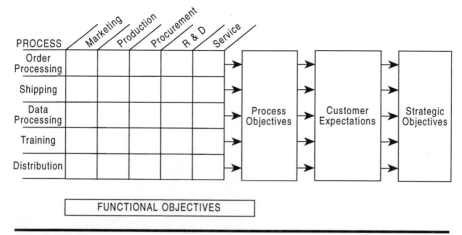

Figure 9-5   Cross-Functional Linkages

The centuries-old hierarchical form of organization with a vertical chain of command was the norm until recently, when organizational complexity demanded horizontal as well as vertical coordination in order to plan and control processes that flowed laterally. If no lateral coordination is achieved, the organization becomes a collection of islands of specialization without integration, a requirement of the systems approach. Linking business process improvement (billing, procurement, recruiting, record keeping, design, sales, etc.) to the key business objectives of the organization is necessary if quality is to become real and relevant. There is widespread agreement that cross-functional teams provide the best vehicle for linking these activities and processes. The concept of linkages is shown in Figure 9-5. Note that a cross-functional approach achieves the objectives of:

■ Customers
■ Functions
■ Processes
■ The organization

Team expert Michael Donovan summarizes a number of trends that will shape the structure and process of employee involvement efforts in the future:[43]

From	To
Perception of employee involvement as a program	Perception as an ongoing process
Voluntary participation	Participation by all members as a natural work team
Quality circles	Several types of teams at many levels
Project focus	Goal focus
Limited management involvement	Active management involvement
Functional management skills	Building participative leadership and facilitation skills into management roles
Employee participation in operating problems	Employee participation in broader issues

QUESTIONS for DISCUSSION

9-1 How does an organizational structure that is focused on classical principles (specialization of labor, unity of command, span of management, delegation of authority) tend to inhibit the implementation of TQM?

9-2 Define the concept of *synergism*. How does organizing around the principles of TQM tend to integrate the organization and achieve synergism?

9-3 What is the concept of the *value chain*? How can it be useful in building an organizational structure?

9-4 In organizing for customer satisfaction, what would be a key activity for
  ■ A brokerage firm
  ■ An aircraft manufacturer
  ■ A retail store

9-5 Explain the concept of the inverted organizational chart.

9-6 Explain how membership in a small group might lead to improved motivation and hence improved quality.

ENDNOTES

1. For example, Michael H. Mescon, Michael Albert, and Franklin Khedouri, *Management,* New York: Harper & Row, 1988, p. 323.
2. W. Edwards Deming, *Out of the Crisis,* Cambridge, Mass.: Massachusetts Institute of Technology, Center for Advanced Engineering Study, 1982, pp. 465–474.
3. Philip B. Crosby, *Quality Is Free,* New York: McGraw-Hill, 1979, pp. 69–70.
4. The Conference Board, *Global Perspectives on Total Quality,* New York: The Conference Board, 1991, p. 9.
5. Michael Porter, *Competitive Advantage: Creating and Sustaining Superior Performance,* New York: The Free Press, 1985.
6. Presentation at the Total Quality Service Management Conference, Dallas, May 21–23, 1990.
7. Kenneth Leach, Vice-President, Administration of Globe Metallurgical, Inc. at the Third Annual Quality Conference, June 22, 1990.
8. Presentation at the Total Quality Service Management Conference, Dallas, May 21–23, 1990.
9. Peter Drucker, *Management: Tasks, Responsibilities, Practices,* New York: Harper & Row, 1974, p. 530.
10. Michael Porter, *Competitive Advantage: Creating and Sustaining Superior Performance,* New York: The Free Press, 1985, p. 144. Signals of value are those factors that buyers use to infer the values a firm creates.
11. Keith Brinksman, "Banking and the Baldrige Award," *Bank Marketing,* April 1991, pp. 30–32.
12. American Society for Quality Control, *'88 Gallup Survey of Consumers' Perceptions Concerning the Quality of American Products and Services,* Milwaukee: ASQC, 1988.
13. Mark Graham Brown, "How to Guarantee Poor Quality Service," *Journal for Quality and Participation,* Dec. 1990, pp. 6–11.
14. In 1981, Cleveland Twist Drill, a Cleveland-based manufacturer of cutting tools with $400 million in sales, had over 500 job classifications in a direct labor force that numbered fewer than indirect labor. Joseph L. Bower et al., *Business Policy,* Homewood, Ill.: Irwin, 1991, p. 588.
15. "How to Profit from Managing Business Process Quality," presentation at the Total Quality Service Management Conference, Dallas, May 21–23, 1990.
16. David Mercer, "Key Quality Issues," in *Global Perspectives on Total Quality,* New York: The Conference Board, 1991, p. 11. Mercer is the project director of the European Council on Quality of The Conference Board Europe.
17. Keith D. Denton, "Horizontal Management," *SAM Advanced Management Journal,* Winter 1991, pp. 35–41.
18. Herbert M. Baum, "White-Collar Quality Comes of Age," *Journal of Business Strategy,* March/April 1990, pp. 34–37.
19. U.S. General Accounting Office, *Quality Management Scoping Study,* Washington, D.C.: General Accounting Office, 1991, p. 23.
20. A company handout entitled "The Motorola Story," written by Bill Smith, Senior Quality Assurance Manager, Communications Sector. The committee's meetings

are described: "The Chief Quality Officer of the corporation opens the meetings with an update on key initiatives of the Quality Program. This includes results of management visits to customers, results of Quality System Reviews (QSR's) of major parts of the company, cost of poor quality reports, supplier–Motorola activity, and a review of quality breakthroughs and shortfalls. This is followed by a report by a major business manager on the current status of his/her particular quality initiative. This covers progress against plans, successes, failures, and what he projects to do to close the gap on deficient results, all pointed at achieving Six Sigma capability by 1992." Discussion follows among the leaders concerning all of these agenda items.

21. Al P. Staneas, "The Metamorphosis of the Quality Function," *Quality Progress,* Nov. 1987, pp. 30–33.

22. There are a number of good sources that provide suggestions for managing change. See, for example, Tom Peters, "Making It Happen," *Journal for Quality and Participation,* March 1989, pp. 6–11; Nina Fishman, "Playing the Transition Game Successfully," *Journal for Quality and Participation,* June 1990, pp. 52–56; John Herzog, "People: The Critical Factor in Managing Change," *Journal of Systems Management,* March 1991, pp. 6–11; Ronald Elliott, "The Challenge of Managing Change," *Personnel Journal,* March 1990, pp. 40–49; Edmund Metz, "Managing Change: Implementing Productivity and Quality Improvements," *National Productivity Review,* Summer 1984, pp. 303–314; and Richard Sparks and James Dorris, "Organizational Transformation," *Advanced Management Journal,* Summer 1990, pp. 13–18.

23. G. Harlan Carothers, Jr., "Future Organizations of Change," *Survey of Business,* Spring 1986, pp. 16–17.

24. Nina Fishman and Lee Kavanaugh, "Searching for Your Missing Quality Link," *Journal for Quality and Participation,* Dec. 1989, pp. 28–32.

25. Nancy Karabatsos, "Quality in Transition: Part One," *Quality Progress,* Dec. 1989, pp. 22–26.

26. David E. Gumpert, "The Joys of Keeping the Company Small," *Harvard Business Review,* July/Aug. 1986, pp. 6–14.

27. With apologies to Maslow and Herzberg, who have provided what is probably the most practical approach to motivation. See Abraham Maslow, "A Theory of Human Motivation," *Psychological Review,* No. 50, 1943, pp. 370–396 and Frederick Herzberg, "One More Time: How Do You Motivate Employees?" *Harvard Business Review,* Jan./Feb. 1968, pp. 56–57. A complete review and summary of the writings of both of these theorists can be found in almost any principles of management textbook. For example, see Michael Mescon, Michael Albert, and Franklin Khedouri, *Management,* New York: Harper & Row, 1988.

28. James H. Harrington, "Worklife in the Year 2000," *Journal for Quality and Participation,* March 1990, pp. 56–57.

29. John Simmons, "Partnering Pulls Everything Together," *Journal for Quality and Participation,* June 1989, pp. 12–16.

30. Rick L. Lansing, "The Power of Teams," *Supervisory Management,* Feb. 1989, pp. 39–43.

31. Larry Gerhard and Walter T. Sparrow, "Pride Teams, A Quality Circle that Works," *Journal for Quality and Participation,* June 1988, pp. 32–36.

32. Vladimir J. Mandl, "Team Up for Performance," *Manufacturing Systems,* June 1990, pp. 34–41.

33. Gary Ferguson, "Printer Incorporates Deming—Reduces Errors, Increases Productivity," *Industrial Engineering,* Aug. 1990, pp. 32–34.

34. In company presentation at the Miami headquarters.

35. As reported in Brian Usilaner and John Leitch, "Miles to Go...Or Unity at Last," *Journal for Quality and Participation,* June 1989, pp. 60–67.

36. Joel E. Ross and William C. Ross, *Japanese Quality Circles and Productivity,* Reston, Va.: Reston Publishing, 1982, p. 6. For those contemplating the establishment of quality circles or other quality improvement teams, this book provides an action plan for the process.

37. James H. Harrington and Wayne S. Rieker, "The End of Slavery: Quality Control Circles," *Journal for Quality and Participation,* March 1988, pp. 16–20. For an example of how the Avco division of Textron revitalized their quality circles with management support, see Peggy S. Tollison, "Managers Are People Too: A Case Study on Developing Middle Management Support," *Quality Circles Journal,* March 1987, pp. 12–15.

38. Rick Lansing, "The Power of Teams," *Supervisory Management,* Feb. 1989, pp. 39–43.

39. Edward E. Lawler and Susan A. Mohrman, "Quality Circles: After the Honeymoon," *Organizational Dynamics,* Spring 1987, pp. 42–54.

40. Carol Gabor, "Special Project Task Teams: An Extension of a Successful Quality Circle Program," *Quality Circles Journal,* Sep. 1986, pp. 40–43.

41. Michael J. Donovan, "Self-Managing Work Teams—Extending the Quality Circle Concept," *Quality Circles Journal,* Sep. 1986, pp. 15–20.

42. U.S. General Accounting Office, *Quality Management Scoping Study,* Washington, D.C.: General Accounting Office, 1991, p. 22.

43. Michael Donovan, "The Future of Excellence and Quality," *Journal for Quality and Participation,* March 1988, pp. 22–24

10

PRODUCTIVITY and QUALITY

In Japan, we are keeping very strong interest to improve quality by use of methods which you started. When we improve quality we also improve productivity.

Dr. Yoshikasu Tsuda
University of Tokyo

During the mid-1980s, the President's Council for Management Improvement wrestled with the productivity process mandated by Ronald Reagan. However, corporate chief executives encouraged the president to get away from processes that stressed productivity and instead to focus on quality. These events led to the creation of the Malcolm Baldrige Award and the subsequent popularity of total quality management (TQM) in U.S. industry.

The relationship among quality, market share, and profitability was examined in Chapter 1, and it was shown that higher quality leads to both increased profits and greater market share. The following questions now arise: Are productivity and quality related? Are they two sides of the same coin? Can you have both? The answer, of course, is *yes*.

Despite a growing body of evidence that indicates a positive correlation, the misconception exists that productivity and cost must be sacrificed if quality is to be improved. In an annual survey of its members in 1990, the Institute of Industrial Engineers (IIE) found the general opinion to be that only when productivity and quality are considered together can competitiveness be enhanced.[1]

There may be some justification for the belief that increased quality means decreased productivity, but it seems to be the view of those who rank production ahead of quality as the top priority. It is argued that a program to improve quality causes disruptions and delays that result in reduced output. While this may be the case in the short run, it generally is not true over a longer time period. As will be discussed in Chapter 11 (The Cost of Quality), such an argument usually fails when the costs associated with poor quality are considered.

The argument for a positive relationship was made by Deming, who based it on the reduced productivity that is caused by quality defects, rework, and scrap. He concluded, "Improvement of quality transfers waste of man-hours and of machine-time into the manufacture of good products and better service."[2] Feigenbaum maintains that a certain "hidden" and non-productive plant exists to rework and repair defects and returns, and if quality is improved, this hidden plant would be available for increased productivity.[3] These arguments are straightforward; any quality improvement that reduces defects is, by definition, an improvement in productivity. The same can be said, of course, for services and for those firms in service businesses. The cost of quality improvement rarely exceeds the savings from increased productivity.

To build a case for or against quality improvement based on output or defect reduction alone is to oversimplify. A more convincing case can be built around the proven benefits of TQM. When the broader picture is considered, it can be shown that increasing quality also increases productivity, and the two are mutually reinforcing.[4] Productivity has come to mean more output for the same or less cost. TQM embraces a broader concept and can be perceived as *including* the benefits of productivity when properly implemented. Productivity has become a tactical short-term approach associated with cost reduction, greater efficiency, better use of resources, and organizational restructuring. TQM is longer term and more comprehensive and as such is concerned with cultural change and creating visions, mission, and values.

Examples of productivity improvements resulting from TQM abound:

■ Under Joseph Juran's guidance, the Internal Revenue Service's processing center in Ogden, Utah adopted quality as a core value, but also achieved productivity increases of $11.3 million from team and management initiatives.

■ NASA's Productivity Improvement and Quality Enhancement (PIQE) program has evolved into a multi-program approach incorporating TQM in the agency and in the contractor work force, which comprises about 60 percent of NASA's total.[5]

▪ The introduction of computer-integrated manufacturing (CIM), combined with TQM and self-directed work teams, resulted in a 50 percent increase in productivity at Monsanto Chemical's Fibers Division.[6]

The LEVERAGE of PRODUCTIVITY and QUALITY

If quality has a leverage effect on market share and profitability (as pointed out here and in Chapter 1), what are the bottom line consequences of productivity improvement?

Confining the illustration to the question of profitability leverage, three hypothetical income statements will demonstrate how small (10 percent) increases in productivity will yield much greater results than a similar increase in sales:

	I Before	II Sales up 10 percent	III Productivity improved 10 percent
Sales	$100	$110	$100
Variable costs	70	77	63
Fixed costs	20	20	20
Profit	$10	$13 (+30%)	$17 (+70%)

In situation I, sales are $100, variable costs $70, and fixed costs $20, yielding a profit of $10. In situation II, a sales increase of 10 percent yields a 30 percent profit increase, while situation III shows a 70 percent profit increase with *no increase in sales*. The leverage is even more dramatic if a smaller and more realistic return on sales is used. There are also potential additional companion benefits that can be achieved in quality. Again, the answer lies in TQM and the continuous improvement of all processes.

MANAGEMENT SYSTEMS vs. TECHNOLOGY

Since the time of Adam Smith's historic 18th century book *The Wealth of Nations,* we have been taught to believe that labor specialization accompanied by mechanization was the answer to economic growth and productivity. The Industrial Revolution proved this to be so. Even today, the conventional wisdom of economists tells us that the rate of productivity growth is largely a function of changes in real capital relative to labor.

There is a continuing debate in Washington regarding the "reindustrialization of U.S. industry" or "supply-side economics" as it is came to be known in the Reagan and Bush administrations. The primary domestic objective of these administrations was the improvement of the productivity of American industry by encouraging greater savings and thus investment in capital stock. Competitiveness, it was said, required an overhaul of U.S. technology. It is generally believed that Japan's quality and productivity advantage comes from advanced technology.

It would be a mistake to attribute Japan's success to technology alone and a bigger mistake to consider technology to be the only answer to improved U.S. quality and productivity. It is not labor replacement that is needed but rather improved processes. Why, for example, would a company invest in advanced computer equipment to improve an information system that is flawed or a manufacturing process that is antiquated? In the first case, the technology will provide bad information more quickly so that poor decisions can be made faster. In the second case, process labor may be replaced only to find an increase in lead time, inventory turn, or cost of quality.

Many people think of technology as automation and mechanization, machines and computers, and semiconductors and new inventions, but the term has a much broader meaning. It is a means of transforming inputs into outputs. Thus, technology includes methods, procedures, and techniques which enable this transformation. It includes both machines and methods. This is worth repeating: technology includes methods that improve processes to improve the output/input ratio. Company after company has achieved remarkable increases in both quality and productivity with little or no investment in the hardware side of technology.

No one can argue convincingly against the use of the hardware side of technology to improve both quality and productivity. The problem is that automation and machines require time and money, both of which are in short supply. Management systems take little of either and may be equally or more effective. The solution is to improve the system—the process—before introducing technology. General Motors has spent more on automation than the gross national product of many countries, yet the excessive cycle time from market research to manufacture resulted in the production of cars that were not competitive. While GM was taking eight years to produce a Saturn, Honda took half as long to market a more competitive car. Honda accomplished this by controlling cycle time and processes.

The general tendency is to focus on technology to reduce labor cost and to overlook the improved quality that can be achieved through improvement of related processes and tapping the potential of the work force. Good companies buy technology to improve processes, reduce lead times, boost quality, and increase flexibility.

Capital spending in service industries has exploded, but there has been very little increase in productivity or quality. Jonathan M. Tisch, president and CEO of Loews Hotels, remarked: "Productivity in manufacturing is advancing five times as fast as in the service sector. In the late 1950s we needed roughly one employee for every four occupied rooms and that was the average across the industry. Today's average, nationwide, is one employee for every two rooms. In other words, productivity is half what it used to be. Despite the advent of the computer and the introduction of many so called labor-saving devices."[7] The focus in both manufacturing and service industries has been on labor productivity, but for most businesses capital intensity does not improve labor productivity enough to keep return-on-investment above the cost of capital. For those businesses that become more capital intensive relative to sales, a decline in return on investment is the result, even if a normal increase in productivity is achieved.[8]

PRODUCTIVITY in the UNITED STATES

The productivity record in the United States is not good. Our capital-intensive industries—home of industrial engineering and the assembly line, production planning, and the computer—have been beaten by Japan and the leading nations of Western Europe as American labor productivity continues to compare unfavorably with the rest of the industrialized world.[9] It is a critical issue for the nation and for individual firms.

Reasons for Slow Growth[10]

When it comes to identifying causes for what has been called the "productivity crisis," every economist, industrialist, and government official seems to have a favorite culprit. Among the most popular explanations are the following issues.

Management inattention. U.S. Secretary of Commerce Malcolm Baldrige (who died in 1987 and for whom the Baldrige Award is named) stated: "Between our own complacency and the rise of management expertise around the world, we now too often do a second-rate job of management, compared to our foreign competitors." One survey by A. T. Kearney, Inc. (management consultants) concluded that the key to productivity is better management and not continued efforts to produce more pounds of automobile per worker. The decade of the 1980s is noted for top management's diversion from productivity, quality, and growth to leveraged buyouts, restructuring, downsizing, and in many case executive perks and golden parachutes.[11]

Short-term gain. The trend has been to focus on short-term financial ratios while failing to take action to ensure long-term growth and productivity.[12] While no one would recommend overlooking financial data, this type of information suffers from the shortcomings of all accounting data. Moreover, financial figures tend to favor the productivity of capital while overlooking the other inputs of labor, material, and energy.

Direct labor. Focus on direct labor has historically been the one variable cost around which financial control systems are designed. Today, the direct labor share of total production costs is down to 8 to 12 percent on average.[13] Some firms fold these costs into overhead or general and administrative expenses, categories that are frequently overlooked when searching for ideas to improve productivity and quality.

Capital. Capital stock formation is largely dependent on savings. Yet Americans appear to be spending more and saving less, leaving fewer dollars for capital formation.[14] The net savings ratios for the major industrialized nations of the world[15] are as follows:

	Average 1980–89	1990	1991
United States	6.0	4.6	4.3
Japan	16.0	14.3	14.5
Germany	12.5	13.4	12.8
Three other major European countries (France, Italy, UK)	14.1	12.0	12.2

Research and development. Expenditures for research and development in the United States surpass every other nation, yet overseas rivals are outpacing the United States in spending growth. Opinion is mixed as to the impact of R&D on productivity. Some evidence indicates that spending is directed toward product improvement rather than productivity improvement. This is good and is expected. However, as previously suggested, many quality investments also improve productivity. An R&D "peace dividend" may be expected from political events in the former Soviet Union and Eastern Europe as R&D dollars move from defense to programs in industry.

Inflation. Is inflation the cause of productivity decline or is inflation the result of the decline?[16] It is almost certain that lower productivity combined with higher wages does result in inflation. To the extent that inflation results in increased relative cost of plant and equipment as compared to labor and the relative cost of operating capital, there can be little doubt that these are investment *disincentives.*

Government regulation, the shift to a service economy, and the lack of goals and programs are among other reasons that have been advanced for the poor record of U.S. productivity. The cumulative effect, although significant, is difficult to estimate.

MEASURING PRODUCTIVITY

Measuring productivity is somewhat easier than measuring quality because the latter is determined by the customer and may be fragmented and elusive. On the other hand, productivity can also be difficult to measure because it is measured by the output of many functions or activities, many of which are also difficult to define.[17] What is the *measurable* output of design, market research, training, or quality assurance?

Despite these difficulties, measures are needed for each activity and in most cases for each individual front-line supervisor. Standards are needed for comparison against past performance, the experience of competitors, and as a basis for action plans to improve.

Carl G. Thor, president of the American Productivity and Quality Center in Houston, is a pioneer in the productivity measurement process and has worked for many years on the development of a measurement system. His principles of measurement for both productivity and quality include:[18]

■ Meet the customer's need—that person who plans to use it. The customer may be external or internal.

■ Emphasize feedback directly to the workers in the process that is being measured.

■ The main performance measure should measure what is important. This may not be the case with the traditional cost control report.

■ Measures should be controllable and understandable by those being measured. This principle may be enhanced by the participation of those being measured.

■ Base measures on available data. If not available, apply cost benefit analysis before generating new data. Information is rarely worth more than the cost of obtaining it.

BASIC MEASURES of PRODUCTIVITY: RATIO of OUTPUT to INPUT

Total factor is the broadest measure of output to input and can be expressed as:

$$\frac{\text{Total output}}{\text{Labor + Materials + Energy + Capital}}$$

This measure is not only concerned with how many units are produced or how many letters are typed, but also considers all aspects of producing goods and services. Hence, this measure is concerned with the efficiency of the entire plant or company.

Partial factor measures are established by developing ratios of total output (e.g., number of automobiles, patients, depositors, students, widgets, etc.) to one or more input categories and are expressed as follows for the partial factor of labor:

$$\frac{\text{Total output}}{\text{Labor input}}$$

The same applies to material, capital, and energy. All measures are ratios of quantities. Although some ratios can be expressed in quantitative terms such as units produced per man-hour, others must combine unlike quantities of inputs, such as tons and gallons of products, employee-hours, pounds, kilowatt hours, etc. To solve this problem, a set of weights representative of the relative importance of the various items can be used to combine unlike quantities. Base period prices are the recommended weights to be used for calculating total productivity, although other weighting systems such as "man-hour equivalents" can be used.

Functional and departmental measures are more likely to benefit the company than an effort to apply comprehensive, company-wide coverage. Most firms rely largely on budgetary dollar accounting data to analyze their operations, even though these data include the effects of inflation, taxes, depreciation, and the arbitrary accounting cost allocations previously mentioned. Because these accounting figures are frequently not significantly related to the activity or process under study, it is desirable to develop measures that reflect output and input in more realistic terms. Where financial measures are used, it is appropriate to deflate them to a base benchmark.

It is important to establish function and activity measures because these organizational entities are where productivity and quality are delivered and where processes are improved. It is here where process design and control happens. A sampling of illustrative measures is provided in Table 10-1.

Individual measures provide the individual supervisor and worker with the basic target for improvement of both quality and productivity through individual action planning. Improvement can only occur if measured against some benchmark (target, yardstick, standard, objective, or result expected).

The simplest and most effective way to set a standard is to list the responsibility of the job on a piece of paper and then list the measures

Table 10-1 Function and Activity Measures

Function/activity	Measure
Customer support	Cost per field technician, cost per warranty callback
Data processing	Operations employees per systems design employees
Quality assurance	Units returned for warranty repair as percentage of units shipped
Order processing	Orders processed per employee, sales per order processing employee
Production control	Order cycle time, inventory turn, machine utilization, total production to production schedule
Shipping	Orders shipped on time, packing expense to total shipping expense
Testing	Man-hours per run-hour, test expense to rework expense

(results expected) that would indicate that the job is being performed satisfactorily. This provides a benchmark from which improvement can proceed. For example:

Responsibility	Measure
Maintenance	Maintain an uptime machine rate of 95 percent
Assembly	Assemble 32 units per man-hour of direct labor
Accounts receivable	Maintain an accounts receivable level of 42 days

Having established these measures, or standards, the individual can then write a *productivity or quality improvement objective* (results expected). Taking the examples above, these improvement objectives could be written:

My productivity (or quality) improvement objective is

Action verb	Results expected	Time	Cost
Improve	Machine uptime from 94 percent to 97 percent	By June 30	At no increase in man-hours or preventive maintenance costs
Increase	Actual production from 90 percent of schedule to 95 percent	Commencing this quarter	At the same cost of manufacture

Industry and competitive measures are important for benchmarking against the competition, best-in-class, and others in the industry. These are examined in Chapter 8 on benchmarking.

Many companies set measures of total factor productivity such as output per labor-hour, material usage rates, ratio of direct to indirect labor, etc., but such macro measures provide little in the way of functional or departmental measures from which an improvement plan can be developed. Unlike return on investment (a measure of capital productivity), which can be broken down into each of its determinants, broad macro measures mean little to those lower in the hierarchy who need specific objectives in order to develop an action plan.

WHITE-COLLAR PRODUCTIVITY

Productivity of white-collar workers is no less important than that of direct labor or manufacturing employees. Indeed, in terms of numbers and expense, staff and non-production employees outnumber production employees by a wide margin. Yet the problem of measurement of output is more elusive. Measuring the units assembled per man-hour is not too difficult, but how many reports should an accountant prepare, not to mention the most difficult of all measures—managerial productivity. Peter Drucker tells us that it is "usually the least known, least analyzed, least managed of all factors of productivity."[19] Research has shown that white-collar employees are productive only about 50 percent of the time. The remainder is non-productive time and can be traced to personal delays (15 percent) and improper management (35 percent). Causes of wasted time include:

Poor scheduling	Poor staffing
Slack start and quit times	Inadequate communication of assignments
Lack of communication between functions	Unproductive meetings and telephone conversations
Information overload	

Measuring the Service Activity

Although the manufacturing worker (one who physically alters the product) has been measured for decades by time standards, time studies, and work sampling, it is not as easy to set standards for the non-manufacturing employee or the service activity. It is unlikely that measurement can be achieved in the same way as is done for the manufacturing worker. Nevertheless, a system can be devised to describe the productivity of an *activity* at a point in time and then provide a baseline for judging continuous improvement over time. The system is particularly appropriate for multi-plant or multi-divisional companies with similar products or services and for individual companies within an industry.

The basis for a system of measurement starts with the existing functions and activities of the organization. Each activity is a subset of a particular function. For example, the *activity* of recruiting is a part of the human resource *function,* accounts receivable is a part of the accounting function, and so on. The typical organization may identify a hundred or more activities that can be grouped into ten or more functions. This concept is shown in Figure 10-1.

The next step is to identify the *output indicators* that "drive" the activities or cause work in the activities. In other words, if it were not

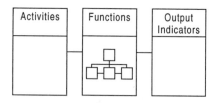

Figure 10-1 Measuring White-Collar (Indirect) Activity

for the work caused by or resulting from the *indicators,* there would be little need for the *activities.* If, for example, there were no personnel employed, there would be no need for employee relations. If there were no purchasing, there would be no need for vendor invoicing. The resources utilized in the activity of vendor invoicing are therefore a dependent variable of the purchasing function. In other words, if activities are the "input" in the productivity ratio of output to input, then the indicators are the "output."

IMPROVING PRODUCTIVITY (and QUALITY)

Improvement means increasing the ratio of the output of goods and services produced divided by the input used to produce them. Hence, the ratio can be increased by either increasing the output, reducing the input, or both. This concept is illustrated in Figure 10-2, along with a sampling of actions and techniques for improving the productivity ratio. This might be called the productivity wheel.

Historically, productivity improvement has focused on technology and capital equipment to reduce the input of labor cost. Improved output was generally thought to be subject to obtaining more production by applying industrial engineering techniques such as methods analysis, work flow, etc. Both of these approaches are still appropriate, but the current trend is toward better use of the potential available through human resources. Each worker can be his or her own industrial engineer—a mini-manager, so to speak. This potential can be tapped by allowing and encouraging people to innovate in one or more of the five ways described in the next section. Employee ideas can improve productivity, and in most cases this is accompanied by an improvement in quality as well.

Figure 10-2   Productivity Wheel

Five Ways to Improve Productivity (and Quality) (see Figure 10-3)

Cost reduction is the traditional and most widely used approach to productivity improvement and is an appropriate route to improvement if implemented correctly. However, many companies maintain a somewhat outdated "across-the-board" mentality that directs each department to "cut costs by 10 percent." Staff services are slashed and training reduced, and the result is an inefficient sales force, reduced advertising, and diminished R&D. Maintenance is delayed and machine downtime

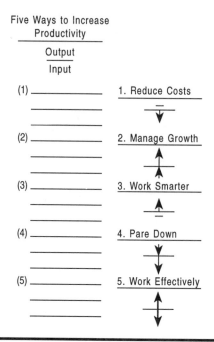

Figure 10-3 Productivity Improvement

is increased. The results may be a non-competitive product and loss of market share.

Under this "management by drive" approach, people are perceived as a direct expense, and the immediate route to cost reduction is seen as cutting this expense as much as possible. This policy usually leads to employee resentment and is frequently counterproductive. It may result in trading today's headache for tomorrow's upset stomach.

Managing growth is a more positive approach, but growth without productivity improvement is *fat*. The improvement may suggest an investment or cost addition, but the investment must return more than the cost, thus increasing the ratio. Capital and technological improvements, systems design, training, organization design, and development are among the many ways to manage growth while improving productivity and quality. The approach does not necessarily mean additional investment in capital improvement. It can also mean reducing the amount of input per unit of output during the growth period. This may be termed *cost avoidance.*

Working smarter means more output from the same input, thus allowing increases in sales or production with the same gross input and lower unit cost. Many companies think that working smarter means putting a "freeze" on budgets while expecting a higher level of output. Although this may be necessary as a stopgap measure, it is hardly a rational course of action to improve productivity over the longer term. Better ways of improving this ratio might be getting more output by reducing manufacturing cost through product design, improving processes, or getting more production from the same level of raw materials by increasing inventory turnover.

Paring down is similar to cost reduction, except that as sales or production is off, input should be reduced by a proportionately larger amount, thus increasing the ratio. This productivity improvement can frequently be achieved through "sloughing off." In many organizations, there are many more opportunities than are generally realized to reduce marginal or unproductive facilities, employees, customers, products, or activities. Peter Drucker puts it this way: "Most plans concern themselves only with the new and additional things that have to be done— new products, new processes, new markets, and so on. But the key to doing something different tomorrow is getting rid of the no-longer-productive, the obsolescent, the obsolete." This "sloughing off" could apply to customers as well. Remember the 80/20 rule.

Working effectively is the best route to productivity and quality improvement; simply stated, you can get more for less. Some ways in which this can be accomplished are suggested in Figure 10-2.

Examples of Increasing Productivity While Improving Quality

Experience has shown that front-line supervisors and employees have a wealth of innovative ideas for productivity and quality improvement. They have only to be asked. In workshops and seminars conducted for hundreds of participants, there has been a high degree of enthusiasm for setting improvement objectives, defining problems, and organizing action plans for improvement. A few that were converted to action plans and resulted in substantial cost reduction as well as improved productivity and quality are presented here as illustrative examples. Each improvement objective will improve the output/input ratio in one or more of the five ways outlined earlier.

■ Improve assembly output by 30 percent by reducing the excessive number and types of fasteners

- Reduce repetitive machine downtime by problem solving
- Set material standards and reduce rework by 10 percent
- Decrease work in process from 45 to 30 days by improved scheduling and shop floor layout
- Improve clerical costs by 30 percent by avoiding duplication with adequate work procedures
- Set standards for setup and improve setup time by 10 percent
- Improve tool revision cost by 50 percent by decreasing lead time from design
- Improve process flow and get 30 percent increased output of presses
- Improve flow of finished goods by improving warehouse layout
- Reduce labor cost by training technicians to replace engineers
- Get more output with less input by cross training and reduction of specialization
- Get more output with same input by better production planning
- Improve bill of materials by reducing custom parts
- Reduce assembly hours by using modular assembly
- Improve reliability by simplified design and design for customer maintainability

CAPITAL EQUIPMENT vs. MANAGEMENT SYSTEMS

Improvements in both productivity and quality have been slowed by two traditional management systems. The first has been the tendency to look to capital equipment as a solution to the problem of labor productivity. In the age of "high-tech," additions to capital have been viewed as the answer to boosting output. There is nothing wrong with this approach. Indeed, as pointed out previously, remarkable gains have been made in mechanization and automation since the Industrial Revolution. However, there are a number of arguments against depending on technology alone. It costs money and takes time, neither of which is an abundant resource.[20] Moreover, direct labor, the focus of capital equipment, is in the range of 8 to 12 percent of total cost of manufacturing. Technology has yet to make significant inroads in the productivity of indirect labor and service industries. Finally, high-tech must be accompanied by low tech—the way workers, supervisors, and managers interact in adapting to new systems.

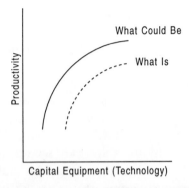

Figure 10-4   Productivity Curve

A basic principle of Economics 101 is illustrated in Figure 10-4. As additional increments of capital are used, productivity increases up to the point where benefits and cost are equal. This is classical economics at its best and reflects Washington thinking about U.S. industrial policy. Figure 10-4 also demonstrates how the productivity curve can be shifted upward by means of improved management systems. This approach costs little and is available immediately. As discussed in earlier chapters, process control and related methods can improve both quality and productivity.

Another shortcoming of the capital investment argument as the primary or sole source of productivity and quality improvement relates to the historical focus on cost reduction. As discussed in Chapter 11, the traditional cost accounting methods of the past provide inadequate information for decision making in the 1990s. Today, decisions on capital expenditures must be based on overall productivity, improving quality, cutting cycle time, reducing inventory, and adding flexibility. Activity analysis is a first step and it is fundamental to improving *management systems.*

ACTIVITY ANALYSIS

Measurement of an activity output is not sufficient. Questions still remain: (1) Is the output/input ratio a positive number? (2) Can this ratio be improved? Most importantly, (3) does the *value added* by the activity contribute to the goal of the organization and the external or internal customer? The overwhelming majority of people in an organization cannot answer either of these questions, except in general and non-measurable terms. They define their activities in terms of what they are doing, not what they want to get done or whether the output is worth more than the input.

People characterized as *input* supervisors or employees are recognized by their dedication to collecting voluminous data for variance reports or closely examining the details of an expense account. The emphasis is on paperwork and the maintenance of records. They are the guardians of company rules and procedures, but are unconcerned about the value of their service to external or internal customers. The means becomes the end. Emphasis is on form and administration (doing things right) rather than process and results (doing the right things). They confuse *efficiency* with *effectiveness*. The design department is efficient at making repeated modifications to the product without regard for the impact on production. The sales force is efficient at calling on the wrong customers with the wrong product. Staff departments are efficient at providing services to internal customers who place no value on the service because they do not have to pay for it. The focus is on the budget rather than results.

Activity-focused supervisors and employees are intent on what they are doing, as opposed to what should be done. The accountant focuses on preparing the cost report rather than reducing overhead costs. The engineer is concerned only with the technical specifications of design without regard to cost, value analysis, or competitive considerations. When asked to define the results of their jobs, these people will reply with such platitudes as "improve the operations," "keep maintenance costs down," or "stay within the budget." It can be said of bureaucracy that focus on activity rather than results seems perfectly logical to those who are trapped within it. The activity may seem logical to the individual performing it, but to an outsider or a customer it is obviously wasteful.

The historical attention that is paid to budgets and cost control has encouraged a focus on activity rather than non-financial measures that plan and monitor sources of competitive value and *strategic* cost information. For most white-collar and service activities, the purpose of the output is to provide input to another downstream activity that can be viewed as the *internal* customer. A good starting point, therefore, is to determine whether the internal customer's expectation is met by the value provided by the upstream activity. The analysis of these activities begins by charting the flow throughout the organization and identifying sources of customer value in each. The central questions to be asked are what is the value added by the activity and what is the output worth to the supplier and receiver.

The major steps in conducting an activity analysis program include:

- Each unit, function, or activity develops a baseline budget that includes a breakdown of one year's costs.
- Set a cost, productivity, or quality target.
- Develop a mission statement for each unit that answers the question: "Why does it exist?"
- Identify each activity that supports the mission and the end products or services that result from that activity.
- Allocate end-product cost that equals the baseline budget.
- Identify receivers (customers) of the end product or service.
- Develop and implement ideas for improvement.[21]

QUESTIONS for DISCUSSION

10-1 Give an example of how improving quality can also increase productivity.

10-2 Illustrate how productivity improvement may be more effective than increased sales in improving profitability.

10-3 How can improved management be as effective as technology and capital equipment in improving productivity?

10-4 Why has the rate of productivity increase been low in the United States?

10-5 Choose four or five functions or activities in staff or white-collar jobs and indicate a measure of productivity for each.

10-6 List three of the five ways to improve the productivity rate of input to output, and identify a specific action that could be taken to achieve the improvement.

ENDNOTES

1. Institute of Industrial Engineers, "Productivity and Quality in the USA Today," *Management Services (UK)*, Jan. 1990, pp. 27–31.

2. W. Edwards Deming, *Quality, Productivity, and Competitive Position*, Cambridge, Mass.: Center for Advanced Engineering Study, Massachusetts Institute of Technology, 1982, pp. 1–2.

3. A. V. Feigenbaum, "Quality and Productivity," *Quality Progress*, Nov. 1977, p. 21.

4. This conclusion is suggested by the Profit Impact of Market Strategies (PIMS) database referred to in Chapter 1. The studies suggest that higher conformance and total quality costs are inversely related and better manufacturing-based quality results in higher output without a corresponding increase in costs. See K. E. Maani, "Productivity and Profitability through Quality: Myth and Reality," *International Journal of Quality & Reliability (UK)*, Vol. 6 Issue 3, 1989, pp. 11–23. One empirical study concludes that improvements in quality level may be related to productivity increases. See Daniel G. Hotard, "Quality and Productivity: An Examination of Some Relationships," *Engineering Management International (Netherlands)*, Jan. 1988, pp. 259–266. In one Conference Board research study of 62 firms that attempted to measure the results of quality on profitability, 47 indicated that profits have increased noticeably because of lower costs and/ or increased market share. See Francis J. Walsh, Jr., *Current Practices In Measuring Quality*, New York: The Conference Board, 1989, p. 3. See also Colin Scurr, "Total Quality Management and Productivity," *Management Services*, Oct. 1991, pp. 28–30.

5. Joyce R. Jarrett, "Long Term Strategy...A Commitment to Excellence," *Journal for Quality and Participation*, July/Aug. 1990, pp. 28–33.

6. Raymond C. Cole and Lee H. Hales, "How Monsanto Justified Automation," *Management Accounting*, Jan. 1992, pp. 39–43.

7. At the Third National Productivity Conference in Dallas on May 21, 1990.

8. Robert D. Buzzell and Bradley T. Gale, *The PIMS Principles*, New York: The Free Press, 1987, pp. 10–11.

9. Slow productivity growth is not characteristic of all U.S. industries. It has been especially high in the manufacture of computers and TV sets, but negative growth has been the case in petroleum refining and retailing.

10. For a more detailed examination of the reasons for slow productivity growth in the United States, see Joel E. Ross *Productivity, People, & Profits*, Englewood Cliffs, N.J.: Prentice-Hall, 1981 and Joel E. Ross and William C. Ross, *Japanese Quality Circles & Productivity*, Englewood Cliffs, N.J.: Prentice-Hall, 1982.

11. For one popular view of the unwillingness of managers to manage, see Robert H. Hayes and William J. Abernathy, "Managing Our Way to Decline," *Harvard Business Review*, July–Aug. 1989, pp. 67–77.

12. "Productivity and Quality in the 90's," *Management Services (UK)*, June 1990, pp. 28–33. This article reports on a survey of British managers, the majority of whom believe that managers are more interested in short-term financial gain than in long-term productivity. Similar surveys in the United States have had similar results.

13. "The Productivity Paradox," *Business Week*, June 6, 1988, p. 103.

14. National Center for Productivity and Quality of Work Life, *Improving Productivity in the Changing World of the 1980s,* Washington, D.C.: U.S. Government Printing Office, 1978.
15. OECD, *Economic Outlook,* July 1991, p. 3.
16. Nobel laureate economist Milton Friedman stated that higher wages and the price-wage spiral are an *effect* of inflation, not a *cause.* Milton Friedman and Rose Friedman, *Free to Choose: A Personal Statement,* Harcourt Brace Jovanovich, 1980.
17. Coopers & Lybrand conducted a survey to determine what federal executives know and think about quality management. About half of the respondents said that the lack of dependable ways to measure quality is a major obstacle. The same could be said of productivity measures. David Carr and Ian Littman, "Quality in the Federal Government," *Quality Progress,* Sep. 1990, pp. 49–52.
18. See Carl G. Thor, "How to Measure Organizational Productivity," *CMA Magazine,* March 1991, pp. 17–19. A company-wide system for measuring productivity is quite complex. The American Productivity and Quality Center conducts a three-day seminar on the topic. See also Brain Maskell, "Performance Measurement for World Class Manufacturing," *Management Accounting (UK),* July/Aug. 1989, pp. 48–50. This article identifies seven common characteristics used by world-class manufacturing firms: (1) performance measures are directly related to the manufacturing strategy, (2) primarily non-financial measures are used, (3) the measures vary among locations, (4) the measures change over time as needs change, (5) the measures are simple and easy to use, (6) the measures provide rapid feedback to operators and managers, and (7) the measures are meant to foster improvement instead of only monitoring.
19. Peter Drucker, *Management,* New York: Harper & Row, 1974, p. 70.
20. Carl Thor, president of the American Productivity and Quality Center in Houston, favors management systems. Regarding high-tech additions, he says: "You need a decade's worth of that kind of investment to have an effect." See a special report entitled "The Productivity Paradox," *Business Week,* June 6, 1988, pp. 100–112.
21. For additional ideas on activity analysis, see Thomas H. Johnson, "Activity-Based Information: A Blueprint for World-Class Management Accounting," *Management Accounting,* June 1988, pp. 23–30. See also Philip Janson and Murray E. Bovarnick, "How to Conduct a Diagnostic Activity Analysis: Five Steps to a More Effective Organization," *National Productivity Review,* Spring 1988, pp. 152–160; Paul L. Brown, "Quality Improvement through Activity Analysis," *Journal of Organizational Behavior Management,* Vol. 10 Issue 1, 1989, pp. 169–179. For a more detailed program for implementing an organization-wide productivity improvement program, see Joel E. Ross, *Productivity, People & Profits,* Englewood Cliffs, N.J.: Prentice-Hall, 1981.

11

The COST of QUALITY

Quality is measured by the cost of quality which is the expense of non conformance—the cost of doing things wrong

Philip Crosby
Quality Is Free

What will it cost to improve quality? What will it cost to not improve quality? These are basic questions that managers need to ask as they focus on the bottom line and company strategic decisions. These questions about the cost of quality have served to draw attention to the quality movement. No one will deny the importance of quality, but it is the confusion surrounding the payoff and the trade-off between cost and quality that is unclear to many decision makers.

It is becoming increasingly clear that whereas the answer to the cost of poor quality may be difficult to obtain, the potential payoff from improvement is extraordinary. Hewlett-Packard estimated that the cost of not doing things right the first time was 25 to 30 percent of revenues. Travelers Insurance Company found that the figure was $1 million per hour. On a positive note, Motorola has reduced the cost of poor quality by about 5 percent of total sales, or about $480 million per year.

COST of QUALITY DEFINED

The cost of quality has been defined in a number of ways, some of which include:

▪ At 3M quality cost equals actual cost minus no failure cost. That is, the cost of quality is the difference between the actual cost of making and selling products and services and the cost if there were no failures during manufacture or use and no possibility of failure.[1]

▪ Quality costs usually are defined as costs incurred because poor quality may or does exist.[2]

▪ The cost of not meeting the customer's requirements—the cost of doing things wrong.[3]

▪ All activities that are carried out that are not needed directly to support departmental [quality] objectives are considered the cost of quality.[4]

These definitions leave unanswered the question: "How much quality is enough?" In theory, the answer is analogous to a principle of economics: basic marginal cost equals marginal revenue (MC = MR). That is, spend on quality improvement until the added profit equals the cost of achieving it. This is not so easy in practice. In economics, the MC and MR curves are difficult to define and more difficult to compute. The same is true of the cost/benefit curves of quality costs. What are the costs of added quality and the "hidden" costs of non-quality? What are the bottom line benefits? Neither of these questions is easy to answer, particularly in view of the long-run strategic implications. The answer lies at the very essence of what the company is about.

The COST of QUALITY

The cost of quality or, more specifically, "non-quality" is a major concern to both national policymakers as well as individual firms. Because much of our national concern with competitiveness seems to be focused on Japan, it is interesting to note that some estimates of quality costs in U.S. firms indicate 25 percent of revenues, while in Japan the figure is less than 5 percent.[5] Estimates of potential savings are as high as $300 billion by nationwide application of total quality management (TQM).[6] Feigenbaum puts the estimate at 7 percent of the gross national product and suggests that this figure can be one of the tools used by policymakers in considering the quality potential of the U.S. economy in relation to the country's major competitors.[7]

The cost of poor quality in individual firms and the potential for improvement can be staggering. In *Thriving on Chaos,* Tom Peters reports that experts agree that poor quality can cost about 25 percent of the personnel and assets in a manufacturing firm and up to 40 percent in a service firm. There appears to be general agreement that the costs range between 20 and 30 percent of sales.[8]

The potential for profit improvement is very substantial. One has only to visualize a profit-and-loss statement with a net profit of 6 percent before tax and then compute what the profit would be if 20 to 30 percent of the operating budget were reduced. Add to this the additional strategic benefits and the potential is great indeed.

THREE VIEWS of QUALITY COSTS

Historically, business managers have assumed that increased quality is accompanied by increased cost; higher quality meant higher cost. This view was questioned by the quality pioneers. Juran examined the economics of quality and concluded that benefits outweighed costs.[9] Feigenbaum introduced "total quality control" and developed the principle that quality is everyone's job, thus expanding the notion of quality cost beyond the manufacturing function.[10] In 1979 Crosby introduced the now popular concept that "quality is free."[11] Today, the view among practitioners seems to fall into one of three categories:[12]

1. *Higher quality means higher cost.* Quality attributes such as performance and features cost more in terms of labor, material, design, and other costly resources. The additional benefits from improved quality do not compensate for the additional expense.

2. *The cost of improving quality is less than the resulting savings.* This view was originally promoted by Deming and is widely held among Japanese manufacturers. The savings result from less rework, scrap, and other *direct* expenses related to defects. This is said to account for the focus on continuous improvement of processes in Japanese firms.

3. *Quality costs are those incurred in excess of those that would have been incurred if the product were built or the service performed exactly right the first time.* This view is held by adherents of the TQM philosophy. Costs include not only those that are direct, but also those resulting from lost customers, lost market share, and

the many hidden costs and foregone opportunities not identified by modern cost accounting systems.

The attention now being given to the more comprehensive view of the cost of poor quality is a fairly recent development. Even today, many companies tend to ignore or downplay this opportunity because of a continuing focus on production volume or frustration with the problem of computing the trade-off between volume and quality. This computational difficulty is compounded by accounting systems that do not recognize the expenses as manageable. More on this will be provided later in this chapter.

One survey of 94 corporate controllers found that only 31 percent of the firms regularly measured costs of quality, and even among those firms productivity was ranked higher than quality as a factor contributing to profit. Not surprisingly, the major reason for failure to measure these costs was lack of top management commitment.[13]

Philip Crosby, of "quality is free" fame, is of the firm opinion that zero defects is the absolute performance standard and the cost of quality is the price of non-conformance against that standard. His concept is catching on as more companies set goals such as parts per million, six sigma, and even zero defects. On the other hand, a goal of zero defects may be more costly than the payoff that might accrue. As one approaches zero defects, costs may begin to increase geometrically.

Another of Crosby's principles, which he calls "absolutes," is *measurement of quality:*

> The measurement of quality is the Price of Nonconformance, not indexes....Measuring quality by calculating the price of waste—wasted time, effort, material—produces a monetary figure that can be used to direct efforts to improve and measure the improvement.[14]

This monetary figure, according to Crosby, is a percentage of sales, and he suggests that the standard should be reduced to about *2 to 3 percent.* This measure has been generally accepted, and many firms use it as a target and measure of progress.

QUALITY COSTS

The costs of quality are generally classified into four categories: (1) prevention, (2) appraisal, (3) internal failure, and (4) external failure.[15]

Prevention costs include those activities which remove and prevent defects from occurring in the production process. Included are such activities as quality planning, production reviews, training, and engineering analysis, which are incurred to ensure that poor quality is not produced. *Appraisal* costs are those costs incurred to identify poor quality products after they occur but before shipment to customers. Inspection activity is an example.

Failure costs are those incurred either during the production process (*internal*) or after the product is shipped (*external*). Internal failure costs include such items as machine downtime, poor quality materials, scrap, and rework. External failure costs include returns and allowances, warranty costs, and the hidden costs of customer dissatisfaction and lost market share. Recognition of the relative importance of external failure costs has caused many companies to broaden their perspective from product quality to total consumer satisfaction as the key quality measure.

In Figure 11-1, the many costs of non-quality are classified into the four categories outlined earlier: (1) prevention, (2) appraisal, (3) internal failure, and (4) external failure. The figure is an attempt to convey the idea of an iceberg, where only 10 percent is visible and 90 percent is hidden from view. The analogy is a good one because the *visible* 10 percent is comprised of such items as scrap, rework, inspection, returns under warranty, and quality assurance costs; for many companies these comprise what they believe to be the total costs. When the *hidden* costs of quality are computed, controlled, and reduced, a firm can acheive the benefits shown at the bottom of Figure 11-1.

Of these types of costs, prevention costs should probably take priority because it is much less costly to prevent a defect than to correct one. The principle is not unlike the traditional medical axiom: "An ounce of prevention is worth a pound of cure." The relationship between these costs is reflected in the 1-10-100 rule depicted in Figure 11-2. One dollar spent on prevention will save $10 on appraisal and $100 on failure costs. As one moves along the stream of events from design to delivery or "dock-to-stock," the cost of errors escalates as failure costs become higher and the payoff from an investment in prevention becomes greater. Computer systems analysts are aware of this and understand that an hour spent on better programming or design can save up to ten hours of system retrofit and redesign. One general manager of Hewlett-Packard's computer systems division observed:

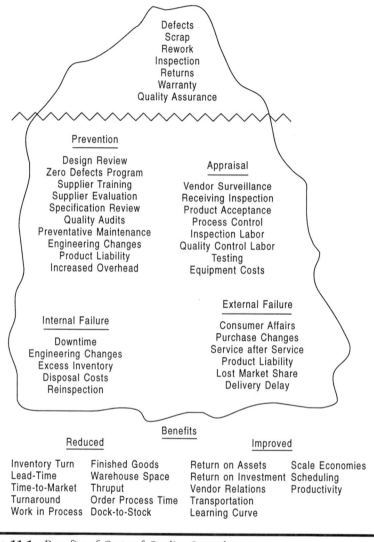

Figure 11-1 Benefits of Costs of Quality Control

The earlier you detect and prevent a defect the more you can save. If you catch a two cent resistor before you use it and throw it away, you lose two cents. If you don't find it until it has been soldered into a computer component, it may cost $10 to repair the part. If you don't catch the component until it is in the computer user's hands, the repair will cost hundreds of dollars. Indeed, if a $5000 computer has to be repaired in the field, the expense may exceed the manufacturing cost.[16]

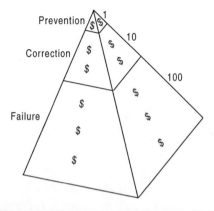

Figure 11-2 1-10-100 Rule

When total customer satisfaction becomes the definition of a quality product or service, it creates a need to develop measures which integrate the customer perspective into a measurement system. This need moves beyond the shop floor and into the many non-product features such as delivery time, responsiveness, billing accuracy, etc. This need also leads to a search for quality, and hence quality costs, in activities not usually recognized as incurring these costs. This will change as more companies realize that all activities can contribute to total customer satisfaction. Thus, quality costs include those factors which lie behind the obvious production processes. Moreover, it becomes necessary to identify the hidden quality costs associated with foregone opportunities.

What is frequently overlooked is the unrealized potential for improved productivity and quality to be achieved by identifying and measuring the difference between no failure (parts per million, six sigma, zero defects, etc.) cost and actual cost. What, for example, would be the payoff from just-in-time, better process control, improved inventory turn, and reduced cycle time in the many cross-functional processes and cost interrelationships in the stream of activities during the life cycle of the product or a service? Each of these actions would improve quality, use fewer resources, and improve return on investment (ROI). How these same actions could also increase market share and profitability was previously examined in Chapter 1. To quote Feigenbaum: "Quality and cost are a sum, not a difference—complementary, not conflicting objectives."[17]

MEASURING QUALITY COSTS

In a 1989 Conference Board survey of 149 large U.S. companies (96 manufacturing), it was found that 111 had a quality process or program. Of the 111 that had a program, 83 attempted to measure quality. The majority of the companies that attempted to measure quality costs compiled the information outside of the accounting system. The breakdown of cost categories reflected a major focus on the direct labor costs of scrap, rework, returns, and costs related to inventory including past-due receivables. There was little evidence to indicate that these costs, once collected, were used to manage processes leading to customer satisfaction quality. Rather, the systems appeared to resemble the traditional cost reduction syndrome discussed in Chapter 10.

An effective cost of quality planning and control systems should be directed toward the basic reason for quality improvement; that is, support of a differentiation strategy. Of course, if a company has not developed a strategy, it becomes difficult to identify those costs of quality that support differentiation of satisfaction in the minds of the customers. For a multi-division or multi-product firm, this strategy may be different for each market segment or strategic business unit. There is little advantage to investing in equipment, overhead, or process improvements which do not add customer value. What is good for Neiman Marcus may not be good for K-Mart.

The cost of differentiation reflects the *cost drivers* of the value activities on which uniqueness is based.[18] Differentiation can also result from the coordination of linked value activities that may not add much cost, but nevertheless provide a cost savings and a competitive edge when integrated.

The measurement and reporting of quality costs to facilitate these strategic demands needs to be provided to users of the information in a form that aids in decision making. Thus, the measurement and reporting of costs of quality should meet the three-part need to: (1) report quality costs, (2) identify activities where involvement is suggested, and (3) indicate interlinking activities.

Activities and functions are not independent. They form a system of interdependencies that are connected by linkages and relationships. For example, purchasing from a low-quality supplier may lead to redesign, rework, scrap, increased field service, and direct labor variance. These linkages are difficult to recognize and are often overlooked. Nor is the conventional accounting system equipped to separate the cost of quality in these linked activities. Virtually all accounting classifications group activities along functional lines and force the reporting of quality costs into several general expense categories such as salaries, depreciation, training, etc. Analyzing the accounts can produce limited estimates of quality costs, but unless the costs are designed into the system, they will be elusive for decisions and action planning.

As one of the steps in the design of a planning and control system, it is useful to identify those activities and linkages between activities where costs occur. Some form of linear or matrix organizational chart or table is useful for this purpose. Departments or activities are listed across the top and costs of quality down the left-hand side. A number (e.g., 1 for primary responsibility or 2 for coordinating responsibility) can be entered at the intersection of the cost of quality category and the activity or function involved. The chart will show overlap among activities and will therefore indicate the need for cooperation, interfunctional teams, and the like. A similar chart can be devised to present cost of quality by activity. Thus, quality costs can be presented based on both cost and activity responsibility, and this form of presentation is more likely to get the attention of top management.

A similar chart can be constructed for reporting the dollar costs of quality. The same format could be used for both budgeting and reporting. Costs can be tabulated by organization unit, by time, by cost of quality categories, or by product. Quality costs can also be normalized for volume by using one or more of the following measures: per direct labor hour, per direct labor cost, per dollar of standard manufacturing cost, per dollar of sales, or per equivalent unit of product.[19]

The most elusive category for reporting is the cost of lost opportunities, which is an external failure cost. These represent the impact of profit from lost revenues resulting from purchase of competitive products and services or from order cancellations due to customer requirements not being met. An additional problem is assigning these *estimated* costs to a quality project or action plan that may prevent recurrence. It is also elusive and difficult to compile the relationships among two or more costs that affect quality costs, i.e., prevention plus appraisal.[20]

The constant theme throughout a cost of quality system is that *costs are not incurred or allocated, but rather are caused.* Cost information does not solve quality problems, nor does it suggest specific solutions. Problems are solved by tracing the *cause* of a quality deficiency.

The USE of QUALITY COST INFORMATION[21]

Quality cost information can be used in a number of ways:

- To identify profit opportunities (every dollar saved goes to the bottom line)
- To make capital budgeting and other investment decisions (quality, as opposed to payback, is the driver of decisions to purchase new equipment or dispose of unneeded ones; equipment for rework is not needed if the rework is eliminated or reduced)
- To improve purchasing and supplier-related costs
- To identify waste in overhead caused by activities not required by the customer
- To identify redundant systems
- To determine whether quality costs are properly distributed
- To establish goals for budgets and profit planning
- To identify quality problems
- As a management tool for comparative measures of input-output relationships (e.g., the cost of a reliability effort versus warranty costs)
- As a tool of Pareto analysis (Chapter 6) to distinguish between the "vital few" and the "trivial many"
- As a strategic management tool to allocate resources for strategy formulation and implementation
- As an objective performance appraisal measure

General Electric's cost of quality system is increasingly emphasizing non-product features such as inquiry responsiveness, delivery times, and billing accuracy. The emphasis is on root cause analysis and process improvement: simplifying procedures and reducing cycle time and driving down quality costs while improving customer satisfaction. Internal and external systems measure performance versus customer expectations; these systems also track opportunities that have been lost by non-conformance to customer expectations.[22]

ACCOUNTING SYSTEMS and QUALITY MANAGEMENT[23]

The shortcomings of accounting information systems were outlined in a previous chapter, and opinions of experts who indicate that accounting information provides little help for reducing costs and improving quality and productivity were reported. The tendency is to *allocate* rather than manage costs. Moreover, the allocation is normally a function of direct labor, an item that has shrunk to 15 percent or less of manufacturing costs. Overhead, at about 55 percent, is spread across all products using the same formula. Accounting also cannot identify or account for the many non-dollar hidden costs of quality and productivity.

Critics claim that management accounting systems should be designed to support the operations and strategy of the company, two dimensions in which quality plays a dominant role. This is increasingly evident in the "new" manufacturing environment, sometimes known as advanced manufacturing technology (AMT), which is characterized by a number of emerging trends. These trends and their implications for quality management were summarized in Chapter 6. Some of the decision-making needs and how traditional accounting practices may fall short in meeting them are listed here:

Decision needs	Traditional accounting
Activity management	Financial accounting
Investment management	Payback or ROI
Non-dollar measures	Dollar accounting
Process control	Cost allocation
Just-in-time	Inventory turn
Feedforward control	Historical control

ACTIVITY-BASED COSTING

The majority of companies that attempt to measure quality costs compile the information and statistics outside of the accounting system. These data are aggregated and do not reflect the true cost of quality or the activity in the process that is causing it. It is worth repeating that costs are not incurred or allocated, *they are caused.* The mere collection of data is of little use unless the data can help identify the drivers of quality costs so that problem identification leads to problem solution.

Activity-based costing (ABC), called "A Bean-Counter's Best Friend" by *Business Week*,[24] can be the system that promises to fill this gap.[25] ABC is a collection of financial and operation performance information that traces the significant activities of a firm to process, product, and quality costs. It is well suited to TQM because it encourages management to analyze activities and determine their value to the customer.

Imagine the case of a firm with excessive warranty costs. The following questions might arise: (1) What is the cost of the returns? (2) What is the cause of the returns and can the cause be traced to a specific activity? Is it the supplier, design, or one of the many activities in production? (3) How can the process(es) be improved to reduce the cost of returns? (4) What is the trade-off between cost of process revision and reduction of warranty costs? (5) What are the strategic implications? The concepts of ABC may lead to some answers.

The concepts of process control and activity analysis were described in Chapter 6 (Management of Process Quality) and Chapter 10 (Productivity and Quality). ABC brings these interlinking concepts together through cross-functional analysis:

■ **Process control** documents the process flow, identifies requirements of internal and external customers, defines outputs of each process step, and determines process input requirements.

■ **Activity analysis** defines each activity within each process and identifies activities as value added or non-value added based on customer requirements.

Activity analysis applies to internal as well as external customers. When Rear Admiral John Kirkpatrick assumed command of the six U.S. Naval Aviation Depots, he inaugurated the use of TQM. One element of the system was that wherever possible, the internal customer was allowed to demand only those internal products or services desired.[26] Could this be a logical extension of customer satisfaction? If it can be applied to external customers, why not internal customers as well?

The third step is to develop cause-and-effect relationships by identifying *drivers* of cost or quality. In the case of cost, the drivers are the conditions that create or "drive" the need for an activity and hence the resources consumed. If the cost driver relates to a non-value activity, it can be eliminated or reduced. It is estimated that 50 percent or more of the activities in most businesses are cost added rather than value added.[27]

ABC recognizes that activities, not products, consume resources, and process value analysis is needed to assign costs to the activities that use them. The system recognizes that costs are driven by factors other than volume or direct labor. In the case of product costing, the costs are assigned based on their consumption of activities such as order preparation, storage time, wait time, internal product movement, field maintenance, and design. The focus on the process, not the product, suggests a transition to breaking down the floor into smaller cost centers and identifying the cost drivers of each.

Cost drivers are agents that cause activity to happen. Consider an engineering change order (ECO) which causes many activities to occur, such as documentation, production schedule changes, purchase of a new machine, or change in a process. If the ECO is issued to correct excessive field maintenance costs, manufacturing will absorb additional charges, marketing's distribution costs will increase, and customer satisfaction may erode because of delays and field repairs. By using the ABC concept, the true cost of the affected product can be determined as well as its cross-functional impact on budgets and performance.

This ECO example illustrates the impact of engineering and design on product life cycle costs. Roughly 80 to 85 percent of a product's lifetime costs, including maintenance and repair expenses, are locked in at this stage. ABC might provide guidelines to help engineers design a product that meets customer expectations and can be produced and supported at a competitive cost.

The Multi-Product Problem

■ At Rockwell International Corporation, a capital budgeting request for an $80,000 laser was denied because at $4,000 per year in labor savings the payback would take 20 years. Further analysis showed that the process would be reduced from 2 weeks to 10 minutes, moving shipments out faster and saving $200,000 a year in inventory holding costs.[28]

■ Tektronix, Inc. adopted ABC in a printed circuit board plant and found that one high-volume product drew on so many resources that it generated a negative margin of 46 percent and sapped profits from other products. These examples illustrate how "across the board" accounting *allocation* of costs, rather than *management* of costs, distorts the information required for good decision making.

There is great potential for inaccurate costing and control of multi-product lines in a firm with a single overhead center, and inaccuracies in costing increase dramatically when allocation is achieved by direct labor, machine time, processing time, or some other "assignment" method. A major soft drink producer found that the costs of its array of brands varied as much as 400 percent from what traditional cost accounting methods reported.

In summary, ABC decomposes activities, identifies the drivers of the activities, and provides measures so that costs can be traced to the activities that cause the cost.

Strategic Planning and Activity-Based Costing

▪ At a meeting of IBM's board of directors in November 1991, various restructuring proposals were considered. One option was to unburden the lines of business from general overhead expenses. For example, the company may remove from its personal computer business the burden of helping pay for research on mainframe computers. (This action was subsequently taken in 1992.)

There is a cost dimension to most strategic decisions. Product lines, channels, locations, brands, segmentation, and differentiation need to be identified, and each decision establishes a linkage between demands and spending on resources. If costs are forecast on the arbitrary basis of some unit directly related to production, the real cost of a product or capital project may be made arbitrarily.[29] ABC can help reveal data for strategic decisions about which product lines to develop or abandon and which prices to increase or decrease. Tracing overhead to activities and then to products may also identify costs that do not contribute to quality and hence to differentiation.

ABC has leapfrogged traditional cost accounting, but it is a new and complicated system. For these reasons, the great majority of companies have not achieved a significant level of sophistication in its use. The basic concept of ABC is that costs of products and quality can be traced to the drivers of activities that consume the resources which *cause* these costs. Research reveals that there is widespread failure to compile the many prevention, appraisal, internal failure, and external failure costs that are "hidden" until identified by a cost of quality management system. If the costs are not identified, there is little chance of tracing them to the process or activity that is causing them. Only the "visible" rework, scrap, and repair/service costs are compiled by more than half of the respondents.

Summary

Is a cost of quality program essential to a quality improvement effort? The answer may be no, but a firm cannot spend unlimited resources without regard for both strategic issues and the cost/benefit equation. Moreover, a cost of quality effort is but one of a system of interlinking efforts that comprise a management philosophy of TQM.

QUESTIONS for DISCUSSION

11-1 Select a firm (restaurant, hotel, airline, manufacturer) and list several costs related to quality failure. Estimate these costs.

11-2 What is the estimated cost of poor quality in U.S. industry?

11-3 What is the justification for Philip Crosby's claim that "Quality Is Free"?

11-4 Illustrate each of the four types of costs of quality.

11-5 Why should prevention costs take precedence over the other three classifications?

11-6 What are the benefits of a cost of quality measuring system?

ENDNOTES

1. Doug Anderson, "How to Use Cost of Quality Data," in *Global Perspectives on Total Quality*, New York: The Conference Board, 1991, p. 37.
2. John F. Towey, "Information Please: What Are Quality Costs?" *Management Accounting*, March 1988, p. 40. Apparently this is a quasi-official definition adopted by the National Association of Accountants (NAA).
3. Roger G. Schroeder, *Operations Management*, New York: McGraw-Hill, 1989, p. 586.
4. J. M. Asher, "Cost of Quality in Service Industries," *International Journal of Quality & Reliability Management (UK)*, Vol. 5 Issue 5, 1988, pp. 38–46.
5. William Band, "Marketers Need to Understand the High Cost of Poor Quality," *Sales & Marketing Management in Canada*, Nov. 1989, pp. 56–59.
6. Ned Hamson, "TQM Can Save Nearly $300 Billion for Nation," *Journal for Quality and Participation*, Dec. 1990, pp. 54–56. This potential is reflected in the quality improvement potential (QIP) index.

7. Armand V. Feigenbaum, "The Criticality of Quality and the Need to Measure It," *Financier,* Oct. 1990, pp. 33–36. This estimate reflects a national (QIP) index for the gross national product. Feigenbaum is president and chief executive officer of General Systems Company, Inc., which installs company-wide quality systems in manufacturing and service organizations.

8. Financial managers estimate the cost at 25 to 30 percent of sales. See Garrett DeYoung, "Does Quality Pay?" *CFO: The Magazine for Chief Financial Officers,* Sep. 1990, pp. 24–34. See also Lester Ravitz, "The Cost of Quality: A Different Approach to Noninterest Expenses," *Financial Manager's Statement,* March/April 1991, pp. 8–13. A 1990 study of quality in North American banks found that non-quality cost related to unnecessary rework and related factors represented 20 to 25 percent of a bank's operating budget. In Britain, the United Kingdom Institute of Management Services estimates that the cost of quality non-conformance amounts to 25 to 30 percent of sales. See John Heap and Lord Chilver, "Total Quality Management," *Management Services (UK),* June 1990, pp. 6–10.

9. J. M. Juran, Ed., *Quality Control Handbook,* New York: McGraw-Hill, 1951.

10. Armand V. Feigenbaum, *Total Quality Control,* New York: McGraw-Hill, 1961.

11. Philip Crosby, *Quality Is Free,* New York: McGraw-Hill, 1979.

12. An excellent discussion of these categories is contained in David A. Garvin, *Managing Quality,* New York: The Free Press, 1988, pp. 78–80.

13. Thomas N. Tyson, "Quality & Profitability: Have Controllers Made the Connection?" *Management Accounting,* Nov. 1987, pp. 38–42.

14. Taken from a promotional brochure by Philip Crosby Associates, Inc. of Winter Park, Florida.

15. The British Science and Engineering Research Council funded a study on quality-related costs as part of a two-year study. A literature review showed the domination of the prevention-appraisal-failure classification, a preoccupation with in-house costs, and little regard for supplier and customer-related costs. See J. J. Plunkett and B. G. Dale, "A Review of the Literature on Quality-Related Costs," *International Journal of Quality & Reliability Management (UK),* Vol. 4 Issue 1, 1987, pp. 40–52. This classification can also apply to non-manufacturing areas. Xerox is one firm that has implemented a well-defined quality program aimed at achieving quality in non-manufacturing services. A model was developed to illustrate the costs of quality based on the prevention-appraisal-failure classification. See Michael Desjardins, "Managing for Quality," *Business Quarterly (Canada),* Autumn 1989, pp. 103–107.

16. As quoted in David A. Garvin, *Managing Quality,* New York: The Free Press, 1988, p. 79.

17. Armand V. Feigenbaum, "Linking Quality Processes to International Leadership," in *Making Total Quality Happen,* New York: The Conference Board, 1990, p. 6.

18. Michael E. Porter, *Competitive Advantage,* New York: The Free Press, 1985, pp. 127–130. Although Porter does not address the specifics of cost of quality, his discussion of differentiation costs provides an excellent dimension to the topic.

19. For more detailed information on methods of compiling quality cost information, see Wayne J. Morse, Harold P. Roth, and Kay M. Poston, *Measuring, Planning, and Controlling Quality Costs,* Montvale, N.J.: National Association of Accountants, 1987. This NAA research study provides a number of actual reporting

formats used by responding companies in the survey. For the collection and reporting formats used by ITT and Xerox, see Francis J. Walsh, Jr., *Current Practices in Measuring Quality,* New York: The Conference Board, 1989 (Conference Board Research Bulletin).

20. James T. Godfrey and William R. Pasewark, "Controlling Quality Costs," *Management Accounting,* March 1988, pp. 48–51.

21. For sources of information regarding the use of cost of quality information, see the following: John F. Towey, "Why Quality Costs Are Important," *Management Accounting,* March 1988, p. 40. See also James M. Reeve, "TQM and Cost Management: New Definitions for Cost Accounting," *Survey of Business,* Summer 1989, pp. 26–30; J. J. Plunkett and B. G. Dale, "A Review of the Literature on Quality-Related Costs," *International Journal of Quality & Reliability Management (UK),* Vol. 4 Issue 1, 1987, pp. 40–52; John J. Heldt, "Quality Pays," *Quality,* Nov. 1988, pp. 26–28.

22. Elyse Allan, "Measuring Quality Costs: A Shifting Perspective," in *Global Perspectives on Total Quality,* New York: The Conference Board, 1991, p. 35 (Conference Board Report Number 958).

23.. H. Thomas Johnson and Robert Kaplan, *Relevance Lost: The Rise and Fall of Management Accounting,* Boston: Harvard Business School Press, 1991. Peat Marwick, one of the Big Six accounting firms, scored a major coup by signing Kaplan to an exclusive contract in the field of activity-based cost accounting. See "A Bean-Counter's Best Friend," *Business Week/Quality 1991* (special bonus issue dated October 25, 1991 entitled "The Quality Imperative: What It Takes to Win in the Global Economy").

24. "A Bean-Counter's Best Friend," *Business Week,* October 25, 1991, pp. 42–43 (special bonus issue entitled "The Quality Imperative: What It Takes to Win in the Global Economy").

25. Because ABC is relatively new, there is no widespread treatment of it in the literature. Perhaps the best source of information, at least from the accountant's point of view, is *Management Accounting.* See, for example, Thomas E. Steimer, "Activity-Based Accounting for Total Quality, *Management Accounting,* Oct. 1990, pp. 39–42. See also Michael R. Ostrenga, "Activities: The Focal Point of Total Cost Management," *Management Accounting,* Feb. 1990, pp. 42–49 and Norm Raffish, "How Much Does that Product Really Cost?" *Management Accounting,* March 1991, pp. 36–39. For a managerial perspective, it is suggested that a literature search be conducted for recent writings of Robert Kaplan.

26. Michael D. Woods, "How We Changed Our Accounting," *Management Accounting,* Feb. 1989, pp. 42–45.

27. Michael J. Stickler, "Going for the Globe. Part II: Eliminating Waste," *Production & Inventory Management Review and APICS News,* Nov. 1989, pp. 32–34.

28. "The Productivity Paradox," *Business Week,* June 6, 1988, p. 104.

29. Bernard C. Reimann, "Robert S. Kaplan: The ABCs of Accounting for Value Creation," *Planning Review,* July/Aug. 1990, pp. 33–34. The author reports Kaplan's contention that the "essence of strategy" is to regard all overhead expenses as variable and driven by something other than the number of units. Also, financial reporting is fine for reporting bottom line financial performance but inadequate for strategic decisions.

II

PROCESSES and QUALITY TOOLS

In this section, the tools and techniques needed to conduct analytic studies for the purpose of quality improvement are discussed. When used within the framework of the Deming cycle (Plan-Do-Check-Act), the techniques can serve as a vehicle for the pursuit of quality.

In Chapter 12, the concept of a process is discussed, and a number of examples of a process are offered. The use of data in a TQM environment is presented in Chapter 13. The two types of data, how to present and describe data, and stratification are among the topics covered in this chapter.

In Chapter 14, the basic quality improvement tools—checksheet, histogram, Pareto analysis, scatter diagram, and cause-and-effect diagram—are presented. Examples are provided to illustrate the use of these basic tools.

In Chapter 15, the concept of process variation, control charts, and how to bring stability to a process are discussed.

12

The CONCEPT of a PROCESS

Everything is a process. Whether it is admitting a patient to a hospital, handling customers at a checkout counter, opening a new account for a bank's customer, or packaging a product for shipment to a customer—all involve a series of activities that are interrelated and must be managed.

WHAT IS a PROCESS?

A *process* is a series of activities or steps used to transform input(s) into output(s). An input or output may exist or occur in the form of data, information, raw material, partially finished units, purchased parts, product or service, or the environment. It is the steps used by an individual or a group to perform work or a complete task. It is sometimes referred to as a technique, method, or procedure. The absence of a clearly defined process makes any activity subject to an arbitrary mode of execution and its outcome or output subject to unpredictable performance. In order to "do it right the first time" and "do the right things right," processes must be effectively managed. When processes are not adequately managed, quality will regress to mediocrity. An organization is a collection of subprocesses. A customer is affected by one or more processes at any given time. Every process has customers (those who depend on it or are affected by it) and suppliers (those who provide the necessary input for that process). Consequently, everyone in an organization serves a customer or serves someone who is serving a customer.

EXAMPLES of PROCESSES

The process of providing higher education is complex. For the sake of simplicity, let us assume that it starts with certain inputs such as SAT scores, a completed application for admission, high school grades, letters of reference, or extra-curricula activities. Several resources are consumed in the process of transforming input to output. The process of transformation includes the activities of teaching, advising, financial aid processing, residence hall assignment and dwelling, library resources, laboratory experience, class group projects, etc. The output is a student who graduates and is competent, drops out, or graduates but is ill-equipped to perform in a competitive work environment. Important feedback can come from industry employers who may criticize the curriculum or simply refuse to hire the graduate. Within this macro process are many subprocesses. One important subprocess is registration. Key inputs would include a list of previous courses taken by the student, the student's classification (freshman, sophomore, junior, or senior standing), the course offerings for that semester, the course prerequisites, etc. Central to this process are academic advising and payment processing (bursar and financial aid). The output is a registered student.

Another well-known process involves packing and shipping a finished product to a customer. The input to the shipping department includes the finished product, the customer's name and address, invoice information, etc. The transformation process entails inspection, making the box, labeling, packing the product, and arranging for shipment. The output in this case is a successful delivery. The feedback loop in this case is the customer reporting back on the condition of the product when it was delivered and whether it was the right product, model, brand, color, quantity, and performance. Everything is a process— whether opening a new account at a bank, taking a customer's order at the drive-through window of a fast-food restaurant, processing an engineering design change, generating a purchase order, or admitting a patient to a hospital.

One of the primary objectives of total quality management (TQM) is to create processes in which individuals or groups will "do it right the first time" and "do the right things right." As suggested in Figure 12-1, individuals or groups can do the right things right or wrong and the wrong things right or wrong. The manner in which individuals do their work (process) can also be right or wrong. The following examples illustrate each of the four quadrants in Figure 12-1.

HOW YOU DO IT

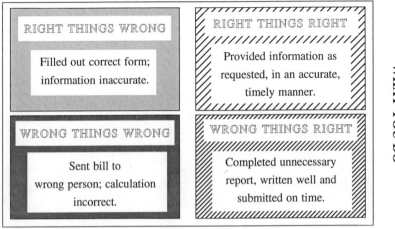

Figure 12-1   The Quality Grid

1. **Doing the right things wrong**
   Examples: (1) You have filled out the correct form, but the information is inaccurate. (2) Using the right equipment but not operating it correctly. (3) A nurse provides the necessary explanation to a sick patient, but in an unprofessional manner.

2. **Doing the wrong things wrong**
   Examples: (1) The accounts department sends an invoice to the wrong customer, and the calculations are incorrect (two processes are affected here—the billing process and the costing process). (2) Filling out the wrong expense reimbursement form and filling it out incorrectly. (3) Picking up the wrong work order and performing the work incorrectly. (4) Purchasing department orders the wrong parts and orders them several weeks late.

3. **Doing the wrong things right**
   Examples: Using the examples in #2 above, (1) the accounts department sends an invoice to the wrong customer, but the calculations are right. (2) Filling out the wrong expense reimbursement form, but filling it out correctly. (3) Violating the optimal job sequence by picking up the wrong work order, but performing the work correctly. (4) Completing an unnecessary report that is well written and submitted on time.

4. **Doing the right things right**
   Examples: (1) Providing the information as requested, in an accurate and timely manner. (2) Ordering the right parts in the right quantity, from the right vendor, within the lead time allowed.

The right things done right means meeting or exceeding the expectations of customers, both internal and external. It also means the elimination of waste, rework, and defects and conformance to valid requirements. If and when it is determined that the customer is incapable of knowing what *right* means, as in the right treatment for a disease, some education and/or explanation would be necessary to bridge the gap between customer expectation and what the service provider delivers.

TYPES of PROCESSES

There are three types of processes, as follows:

▪ **Management process:** This entails the method(s) used by management in executing their management functions. Three key functional areas used by management are planning, organizing, and controlling.

▪ **Functional process:** A functional process consists of the methods used to achieve functional objectives within a group or by an individual.

▪ **Cross-functional process:** This includes the method(s) used to achieve objectives that require participation or input from more than one group or individual. For example, the problem of minimizing breakage of a fragile product might require input from the shipping, design, marketing, packaging, and manufacturing departments. Similarly, the problem of an adverse drug reaction in a hospital may require the involvement of the pharmacist, the ordering physician, a registered nurse, and a unit secretary. Each group or individual controls one or more of the subprocesses affecting the problem.

The TOTAL PROCESS

TQM calls for an evaluation of the total system, not just the subsystems. The danger of suboptimization always exists in that subprocesses instead of the total process are optimized. In many instances, the total process is not defined, and therefore, accountability is conspicuously absent. Each internal customer provides intermediate inputs or receives intermediate outputs throughout the process. These intermediate inputs and outputs are used to achieve the final outcome of the organization. The external customers provide an initial input to or receive final output from the process. Because everyone in an organization serves a customer or serves someone who does, everyone is, therefore, part of a customer–supplier chain. No worker's task is isolated. Consequently, no worker is expected to either accept or pass on defective work or product.

The Feedback Loop

The need to constantly improve processes makes it imperative that a feedback loop be introduced into every process. This feedback loop becomes the link between the output or outcome and the input. It provides the system with an opportunity to evaluate the gap between the expectation of the customer (internal or external) and what is produced or delivered by the supplier. The real value of feedback lies in its usefulness in analyzing the process of transformation.

QUESTIONS for DISCUSSION

12-1 What is a process?
- Give an example of a macro process in either service or manufacturing.
- Give examples of at least two subprocesses within the macro process defined above.

12-2 Define the feedback loop for the macro process and the subprocesses in Exercise 12-1. How can feedback be used to improve the processes you have identified?

12-3 What are the three types of processes? Give examples to illustrate each type.

12.4 Give an example of each of the following in a service and manufacturing sector:

- Doing the right things wrong
- Doing the wrong things wrong
- Doing the wrong things right
- Doing the right things right

ENDNOTES

1. H. Gitlow, S. Gitlow, A. Oppenheim, and R. Oppenheim, *Tools and Methods for the Improvement of Quality,* Homewood, Ill.: Irwin Publishers, 1989.
2. V. K. Omachonu, Total *Quality and Productivity Management in Health Care Organizations,* Milwaukee: Quality Press, American Society for Quality Control, and Norcross, Ga.: Industrial Engineering and Management Press, Institute of Industrial Engineers, 1991.

13

TQM and DATA

TYPES of DATA

Facts concerning a process, service, product, person, or machine are considered data. Data can be categorized into two types: attribute and variable. An attribute refers to a quality characteristic of a product, process, etc. that can be counted. **Attribute data** is therefore referred to as countable or discrete data. It commonly follows yes/no or go/no-go criteria. Examples of attribute data are provided in Table 13-1.

Note that in each of the cases listed in Table 13-1, the criteria must be well defined. For example, is a late payment defined as the difference between the due date and the date of actual payment? If it is, then the number of times payments were made after due date is being counted. Similarly, in determining the number of times a receptionist fails to answer the phone on time, it is important to first establish what "on time" means. Is it, for example, calls answered within three rings? Such a standard must first be established prior to data collection.

Variables data involves measurement to reflect criteria such as quantity, size, or length. This type of data exists on a scale that can be divided into an infinite number of increments, and thus it is continuous data. The following are examples of variable data:

- The time it takes to assemble a chair
- The diameter of a shaft
- Response time to a patient's call for help
- Average food temperature
- Customer waiting time

Table 13-1 Examples of Attribute Data

Attribute	Criteria
Number of late payments	Whether or not a payment is late
Number of defective pens	Whether or not a pen writes (pass/fail)
Percent idleness of a machine	Whether or not a machine was idle
Number of medication errors	Whether or not a medication was given in error
Number of rings before phone is answered	Whether or not phone was answered in time

The computation of a value such as miles per gallon is considered variable data because it is derived from a measured value (the number of miles driven and the number of gallons of gasoline in the car). In the same way, any computation, such as a proportion or ratio, that comes from attributes data is itself considered attribute data. Therefore, in general, the data type can be determined from the original data.

HOW to PRESENT/DESCRIBE DATA

Visual Description: Tabular Displays

Data can be described in tabular form by means of frequency distributions or cumulative frequency distributions. A frequency distribution shows the frequency, or number of times, a given value or values occur. An example of such a tabulation and the information it provides is given in Figure 13-1.

The absolute frequency column denotes the number of occurrences in each category, e.g., 57 of the 90 wheelchairs in the sample had no

# of defects	Absolute frequency	Relative frequency (%)
0	57	63.33
1	20	22.2
2	7	7.78
3	6	6.67

Figure 13-1 Frequency Distribution of the Number of Defects in a Sample of 90 Wheelchairs

Battery life (hr) less than or equal to	Absolute frequency	Cumulative frequency	Relative cumulative frequency (%)
50	25	25	42.67
100	10	35	58.33
150	11	46	76.67
200	14	60	100

Figure 13-2   Cumulative Frequency Distribution of the Life of Batteries from a Sample of 60 Batteries

defects. The relative frequency refers to the number of occurrences in a particular category relative to the total occurrences, given as a percentage. For example, 6.67 percent (or 6/90 × 100) of the wheelchairs sampled had three defects.

The cumulative frequency distribution is used, for example, when the interest lies not in how many occurrences/items exist in one category, but rather in the frequency of occurrences/items with a value less than some measurement. For example, how may batteries lasted less than 50 hours, how many lasted less than 100 hours, and so on. The cumulative frequency distribution for this example is shown in Figure 13-2. From this figure, one readily knows that 58.33 percent of the batteries last 100 hours or less and 76.67 percent of the batteries last 150 hours or less.

Tabular displays of data are very useful in providing frequency information. It is important, however, to note that these displays do not provide any insight as to how the data were collected or any possible trends that may exist in the data.

Visual Description: Graphical Displays

Data are often presented graphically in order to illustrate relationships and trends that are not visible in tabular displays. Variables data are usually presented in histograms. Attributes data are usually presented in bar charts, although histograms can be used for either type of data. Bar charts are very similar to histograms except that the bars do not enclose an interval. Instead, they are centered about each of the attribute categories. These two types of graphs are illustrated in Figures

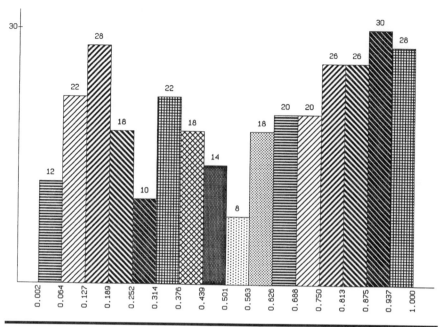

Figure 13-3   Histogram

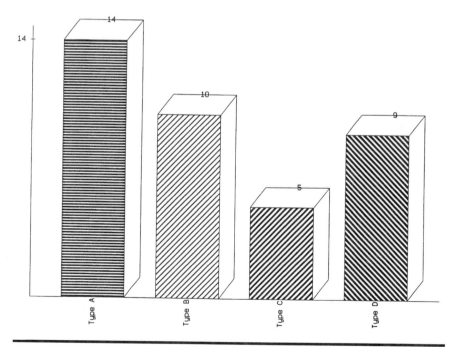

Figure 13-4   Bar Chart

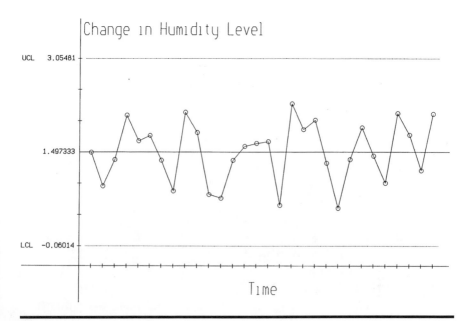

Figure 13-5 Run Chart

13-3 and 13-4. The methodology for constructing a histogram will be explained later.

A run chart (Figure 13-5) is another important tool for displaying data graphically. It is a graph that plots the values of the characteristic being studied versus time, thereby allowing the detection of trends over time.

Numerical Description

To numerically describe data, two aspects or properties are usually presented: the central tendency and the dispersion. Central tendency refers to the central portion of the data, while dispersion refers to the spread of the data.

The main measures of central tendency are mean, median, and mode. The **mean** is simply the arithmetic average of the values. It is denoted by $\overline{X}$ (Xbar) and is calculated by

$$\overline{X} = \frac{\Sigma X}{n} \tag{13-1}$$

where X is the data value and n is the total number of observations.

Another way to state this is the sum of X (ΣX) divided by the number of observations.

Consider the example in which a refrigerator manufacturer provides the time (in minutes) for packing nine refrigerators::

$$10 \quad 18 \quad 12 \quad 17 \quad 19 \quad 16 \quad 12 \quad 11 \quad 14$$

The mean may be calculated using Equation 13-1 as follows:

$$\overline{X} = \frac{10 + 18 + 12 + 17 + 19 + 16 + 12 + 11 + 14}{9}$$

$$= 14.3 \text{ minutes/refrigerator}$$

The mean is the most common of the three measures of central tendency. It is mathematically the strongest because it takes into account every data point.[1]

The **median** is the middle value when the data are arranged in ascending order. Its advantage over the mean lies in the fact that it is less affected by the extreme values. Using the previous data set for refrigerator packing times, the data are arranged in ascending order as follows:

$$10 \quad 11 \quad 12 \quad 12 \quad \mathbf{14} \quad 16 \quad 17 \quad 18 \quad 19$$

$$\text{median} = \mathbf{14}$$

In this data set, the number of observations was odd, which made it easier to find the median. Had the number of observations been even, the median would have been calculated by taking the mean of the middle two values. Usually, however, the median is used when there is an odd number of observations.

The third measure of central tendency is the **mode**. It is the value that occurs most frequently in the data set. Again, following the previous example, the mode is identified as follows:

$$10 \quad 18 \quad \mathbf{12} \quad 17 \quad 19 \quad 16 \quad \mathbf{12} \quad 11 \quad 14$$

$$\text{mode} = \mathbf{12}$$

In this example, there is only one mode, but this is not always the case. A data set with two modes is regarded as bimodal, and one with more than two is known as multimodal. The mode can also be found by creating a frequency table and choosing the value(s) with the highest frequency.

The dispersion or variability of the data is measured by the range and the standard deviation. The **range**, R, measures the spread of the data by subtracting the lowest value from the highest value. In this data set, the range is

$$R = 19 - 10 = 9$$

Therefore, the data in this example have a spread of 9 minutes.

The **standard deviation** is another measure of spread, but it takes into account all of the data. The sample standard deviation, S, can be calculated as follows:

$$S = \left[\frac{\Sigma(X - \bar{X})^2}{n - 1} \right]^{1/2} \tag{13-2}$$

1. The mean, $\bar{X}$, must first be calculated using Equation 13-1. From the previous calculation, we know the mean to be $X = 14.3$

2. $(X - \bar{X})$ is calculated for each reading X and is then squared:

$$10 - 14.3 = -4.3$$
$$-4.3^2 = 18.49$$

Similarly,

$$18 - 14.3 = 3.7$$
$$3.7^2 = 13.69$$

and so on.

3. These numbers are then added:

$$18.49 + 3.69 + 5.29 + 7.29 + 22.09$$
$$+ 2.89 + 5.29 + 10.89 + 0.09 = 86.01$$

4. This value is then divided by the number of observations minus one ($n - 1$):

$$\frac{86.01}{9 - 1} = 10.75$$

This value is called the variance.

5. Finally, taking the square root of the variance gives the standard deviation of the sample as per Equation 13-2:

$$(10.75)^{1/2} = 3.28$$

Note: Data are obtained from a sample. A distinction must be made between a sample and a population. For example, values for the time it takes a few account representatives to open new checking accounts for first-time customers constitute a sample and yield a certain mean and standard deviation. These values, however, differ from the true mean and standard deviation, which would have to be calculated from the population. In this case, the population would be all account representatives who open new checking accounts, rather than just a group of them.

This difference between sample and population is reflected in the notation and calculation of the mean and standard deviation. The true (population) parameters are denoted by Greek letters: μ (mu) denotes the mean and σ (sigma) the standard deviation. For both samples and the population, the mean is still simply an arithmetic average, but the notation differs (μ and σ). The standard deviation of the population is calculated as follows:

$$\sigma = \left[\frac{\Sigma(X - \overline{X})^2}{n} \right]^{1/2} \qquad (13\text{-}3)$$

When the true parameters μ and σ are not known, X and S are used as estimators of these parameters. This is denoted by a circumflex ($^\wedge$) over the parameters μ and σ so that $\hat{\mu} = X$ and $\hat{\sigma} = S$.

Stratification of Data

The stratification of data involves breaking up the data into smaller related subgroups, so as to make the analysis clearer and more precise. The following examples illustrate the application of the concept of stratification:

■ The XYZ Corporation manufactures six product lines and distributes them to four regions of the country. The company utilizes the services of three different distribution outfits for moving its products to its customers. During the last year, the company has experienced a tremendous increase in the number of customer complaints regarding damaged products. The company recorded a total of 3000 complaints from its customers during the year. Every customer who receives a damaged product registers a complaint by returning the merchandise.

In order to understand the nature of the problem, it is essential to stratify the available data. The data could be broken down according to the following criteria:

1. Complaints by region
2. Complaints by distribution outfit
3. Complaints by product line

Depending on the type of problem and the type of data available, other forms of stratification may be pertinent. Other forms may include stratification by day of the week (Mondays and Fridays are important days of the week for certain types of customer-related problems). Other examples include stratification by time of day, type of defect, age group, skill type, price range, noise level, work center, and weight.

■ A midwestern airline is interested in investigating its high rate of missing luggage. It recorded a total of 600 missing pieces of luggage during a one-year period. This accounts for about 8 percent of its total pieces of luggage handled during the period. In applying the concept of stratification, the airline may wish to break down the data into categories and the data could be stratified by:

1. Airport type (small, medium, and large airports)
2. Direct flight or transfer
3. Shift (morning, afternoon, and evening)
4. Economy or first-class passengers
5. Day of the week
6. Type of luggage

A major advantage of stratification is that it helps the analyst understand the context in which the problem occurs.

QUESTIONS for DISCUSSION

13-1 What are the two types of data described in the chapter? Give three examples of each.

13-2 What two properties of data are necessary to numerically describe data?

13-3 The following data represent the weekly number of complaints recorded for a particular product over a 9-week period: 8, 20, 14, 19, 17, 14, 8, 10, and 6.
- What is the mean of the data?
- What is the median?
- What is the mode?

13-4 How is the dispersion or variability of the data given in Exercise 13-3 measured?
- Determine the range of the data.
- What is the standard deviation?

13-5 How would you apply the concept of stratification of data to the following situations:
- Payroll errors occurring in a major utility company
- Complaints regarding food services at a hospital cafeteria
- High drop-out rate among students in a certain major at a university

ENDNOTES

1. H. Gitlow, S. Gitlow, A. Oppenheim, and R. Oppenheim, *Tools and Methods for the Improvement of Quality*, Homewood, Ill.: Irwin Publishers, 1989.
2. V. K. Omachonu, *Total Quality and Productivity Management in Health Care Organizations*, Milwaukee: Quality Press, American Society for Quality Control, and Norcross, Ga.: Industrial Engineering and Management Press, Institute of Industrial Engineers, 1991.

14

QUALITY IMPROVEMENT
TOOLS

The Deming cycle, also known as the Plan-Do-Check-Act (PDCA) cycle, is a systematic approach aimed at helping management with continuous quality improvement. The cycle involves developing a plan for improvement, implementing (doing) the plan, checking the results, and acting based on those results. Then the cycle begins again with revision of the plan. The process of quality improvement is thus continuous. The Deming cycle is illustrated in Figure 14-1.

Planning for improvement entails knowing what the customer wants and needs; the process goals are determined from this information. Examining the current process with respect to these goals will help management develop a plan for improvement. The *plan* stage requires defining the problem, some data collection and analysis, searching for possible solutions, evaluating alternative solutions, and a recommendation for action.

The *do* stage follows the implementation of this plan. The plan is implemented as a test case. This may be done in a controlled setting and on a small scale, from which data must be gathered.

In the *check* stage, the results of the *do* stage are analyzed to determine whether the desired objectives are being met. Based on this analysis, the *act* phase begins. If the results are positive, action is then taken to replicate the improvement across the board. It is necessary at this stage to standardize and to preserve the gains made. If the results are negative, the plan stage is revisited and the cycle begins again. Again, the cycle is repeated endlessly.

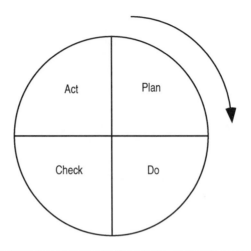

Figure 14-1 Deming Cycle

Within the stages of the Deming cycle, other tools are used to aid in organizing data in order to extract information. These quality control tools include check sheets, histograms, Pareto charts, cause-and-effect diagrams, scatter diagrams, and control charts.

CHECK SHEET

A check sheet provides a method for viewing data (attribute or variable) as it is being collected. It is a form designed to collect data in a systematic and consistent manner. The most common type of check sheet consists of checking the appropriate interval in which an observation falls as it is being observed. As the number of checks increases, the distribution of the data becomes evident.

Example

The check sheet in Figure 14-2 shows attribute data collected on the number of phone calls answered by various receptionists throughout a company on a particular day. When using variable data, the categories simply become intervals, e.g., between 10 and 15 seconds, between 15 and 20 seconds, and so on.

Receptionists	Talleys
A	//// //// //
B	//// //// //// /
C	//// //// //// //// //// //
D	//// //// ///
E	//// //

Figure 14-2   Check Sheet

HISTOGRAM

A histogram is a tool which plots the frequency of data against the values of the data grouped in intervals. This allows the data to be displayed by central tendency and spread and thus makes it possible to determine what distribution it follows. Three aspects of the intervals must be established: number of intervals, the range, and where the center is. These parameters generally depend on the number of data points available. There is no one method to determine these values. In fact, histograms frequently must be plotted more than once, in different ways, until the underlying distribution becomes evident. In all cases, however, a large number of data points (observations) is preferable. The number of intervals can be determined by using a guide similar to Table 14-1 or by using formulas such as one developed by Ishikawa:

$$C = 6 + \frac{N}{50}$$

where C = the number of class intervals and N = the number of observations.

Table 14-1   Histogram Guide

# of observations	# of intervals
<50	5–7
50–100	6–10
100–250	7–12
>250	10–20

Source: Banks.[1]

Table 14-2 Data Sheet for Histogram

Customer arrival times					Class intervals	Observations per interval
11:00	11:25	12:03	12:11	12:25		
11:00	11:29	12:03	12:12	12:25	11:00–11:17	8
11:02	11:31	12:03	12:12	12:30	11:18–11:35	5
11:05	11:45	12:05	12:12	12:30	11:36–11:53	3
11:07	11:49	12:06	12:14	12:31	11:54–12:11	15
11:12	11:49	12:06	12:16	12:33	12:12–12:29	11
11:13	11:56	12:06	12:17	12:35	12:30–12:47	7
11:15	11:59	12:06	12:17	12:40	12:48–1:00	1
11:20	12:00	12:08	12:17	12:44		
11:20	12:00	12:09	12:17	12:57		

Example

The number of people visiting a fast food restaurant at lunch time was recorded for the purpose of scheduling employees. Data were recorded for the hours between 11 a.m. and 1 p.m. The establishment was visited by 45 customers during this time interval. According to Ishikawa's equation, the number of intervals should be

$$C = 6 + \frac{45}{50} = 6.9 \rightarrow 7$$

Thus, the data should be divided into seven classes, as displayed in Table 14-2. The histogram, then, is as illustrated in Figure 14-3.

PARETO ANALYSIS

Pareto analysis provides one of the most powerful ways of looking at data for the purpose of quality improvement. It is based on the 80/20 rule: 80 percent of the problems can be attributed to 20 percent of the causes. Hence, the Pareto chart identifies the cause of the majority of the problems: the vital few as distinct from the trivial many. It is constructed as demonstrated in the following example.

Sixty-one customers reported being dissatisfied with the assembly work required to put together a new brand of children's bicycles. The

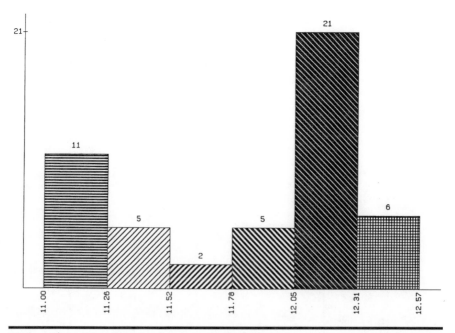

Figure 14-3 Histogram for Customer Arrival Times

reasons for their dissatisfaction are presented in Table 14-3. *Note:* The category "other" always appears last, because it encompasses various types of complaints, each with fewer occurrences than the smallest individual complaint. The category "other" cannot be greater than one-half of the largest category.

Table 14-3 Data Sheet for Pareto Diagram

Types of complaints	Number of complaints	Cumulative total	Percent of overall total	Cumulative percent
Insufficient drawings in instruction manual	33	33	54	54
Complex tools required	12	45	20	74
Instructions in English only	4	49	7	81
Wording is ambiguous	2	51	3	84
Other	10	61	16	100
Total	61	—	100	—

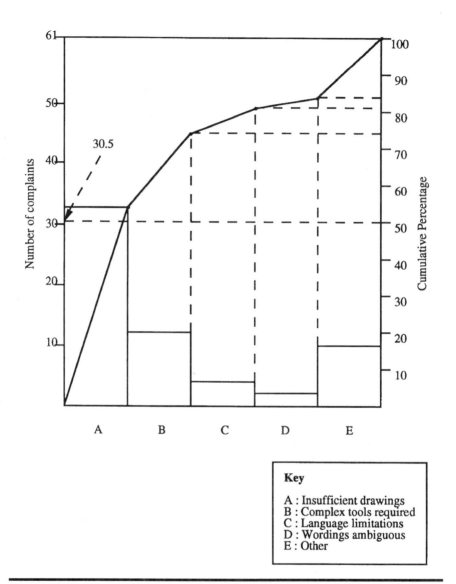

Figure 14-4 Pareto Diagram by Types of Complaints

The graph has two vertical axes. One ranges from 0 to the total number of observations, and the other has a scale from 0 to 100 percent. The horizontal axis is divided into the number of categories, which in this example is types of complaints. The cumulative curve is then plotted by marking the cumulative points above the top right-hand corner of each bar (interval). Refer to Figure 14-4 for the Pareto chart for the bicycle assembly example.

SCATTER DIAGRAMS

A scatter diagram is used to investigate the relationship between any two characteristics, for example, the number of people absent from work and the number of complaints from customers on the same day. A scatter diagram plots one characteristic against another in order to identify whether a relationship exists between them. Two characteristics are chosen. The dependent variable is listed on the vertical axis, with its appropriate scale. Scales should be constructed so that the x and y axes are approximately equal in length. The data (x,y) are then plotted. Interpretation of the diagram is based on the appearance of the marks. Positive correlation means that as x increases, y increases; negative correlation means that as x increases, y decreases.

Example

The relationship between a patient's length of stay (LOS) (x) at a hospital and level of satisfaction (y) as rated by the patient on a scale of 1 to 10 (10 = highest level of satisfaction) is listed in Table 14-4. The scatter diagram for these data is provided in Figure 14-5.

Based on this hypothetical scenario, the two factors seem to be negatively correlated, which means that the longer a patient stays in the hospital, the less satisfied he or she will be in general.

Although scatter diagrams are usually interpreted visually, they can be more precisely interpreted by calculating the correlation coefficient, r. This coefficient ranges from −1 to +1, with −1 signifying a perfect negative correlation (straight line) and +1 signifying a perfect positive correlation. Zero, then, represents no correlation. The coefficient is calculated using the following formula:

$$ r = \frac{n\Sigma(xy) - \Sigma x \Sigma y}{[(n\Sigma(x^2) - (\Sigma x)^2)(n\Sigma(y^2) - (\Sigma y)^2)]^{1/2}} $$

where n = sample size, y = a dependent variable, and x = an independent variable.

Table 14-4 Effect of Length of Stay (LOS) in Hospital on Patient Satisfaction

Patient	Satisfaction level	LOS (days)	xy	y^2	x^2
1	3	10	30	9	100
2	6	4	24	36	16
3	4	11	44	16	121
4	4	10	40	16	100
5	9	3	27	81	9
6	7	3	21	49	9
7	10	1	10	100	1
8	4	6	24	16	36
9	5	7	35	25	49
10	5	6	30	25	36
11	7	4	28	49	16
12	5	5	25	25	25
13	3	8	24	9	64
14	5	10	50	25	100
15	3	12	36	9	144
16	3	10	30	9	100
17	2	14	28	4	196
18	8	2	16	64	4
19	8	1	8	64	1
20	1	14	14	1	196
21	3	9	27	9	81
22	5	7	35	25	49
23	6	6	36	36	36
24	7	6	42	49	36
25	4	9	36	16	81
26	4	10	40	16	100
27	7	7	49	49	49
28	6	5	30	36	25
29	2	11	22	4	121
30	9	4	36	81	16
Total (Σ)	155	215	897	953	1917

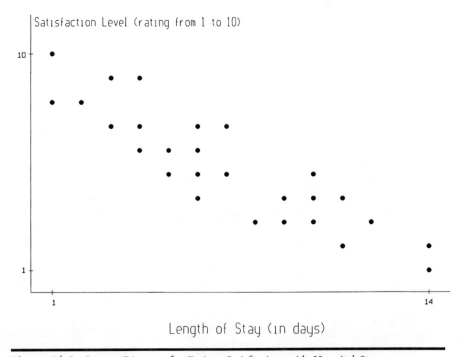

Figure 14-5   Scatter Diagram for Patient Satisfaction with Hospital Stay

Using the values in the previous example, the correlation coefficient between LOS and the patient's level of satisfaction is calculated:

$$r = \frac{30(897) - (255)(115)}{[(30(1917) - (215)^2)(30(953) - (155)^2)]^{1/2}}$$

$$r = \frac{-6415}{[(11285)(4565)]^{1/2}}$$

$$r = \frac{-6415}{7177.47}$$

$$r = -0.89$$

This value indicates a strong negative correlation between the two factors.

CAUSE-and-EFFECT DIAGRAM

A cause-and-effect diagram, also known as an Ishikawa or fishbone diagram, shows the relationship between a quality characteristic and certain factors. It is a graphic tool which shows the relationship between causes and effects. Its principal purpose is to help identify the root cause of a problem. The construction of a cause-and-effect diagram begins with brainstorming, during which all factors affecting the quality characteristic or problem being considered are listed. Once these factors are listed and developed, they are then categorized. The actual diagramming then begins. (See Figure 14-6 for an example of the application of a cause-and-effect diagram to the problem of delays in registration.) The process is as follows:

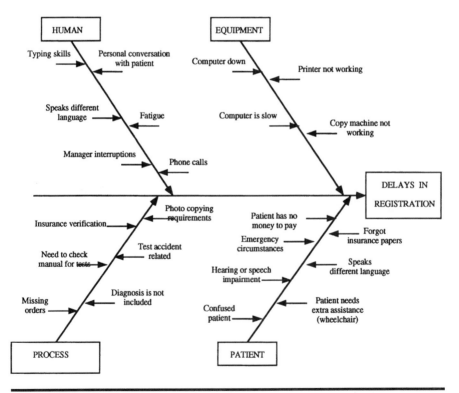

Figure 14-6 Cause-and-Effect Diagram for Delays in Registering a Hospital Patient

1. The effect or problem is placed in a box at the right side of the page.

2. A horizontal line is drawn (from left to right) to the box. Other lines stem from this "backbone," linking to the primary causes. Lines for the secondary causes stem from these lines, and so on for tertiary causes. All lines flow toward the effect as denoted by arrowheads.[2]

3. Determine categories of causes for the effect. Generic categories frequently used include *human, methods, materials, equipment,* and *environment.* Other categories that are more pertinent to the process being studied may also be used. Categories may be determined by breaking down the process into phases.

4. Draw diagonal lines above and below the horizontal line and label with the categories.

5. Generate a list of causes for each category (usually from the brainstorming list).

6. Arrange the causes on each bone, drawing branch bones to show relationships among the causes.

7. Generate "root causes" by asking "why?" until a useful level of detail is reached. Be sure to distinguish between a cause and a symptom.

The cause-and-effect diagram is helpful in categorizing the causes of a problem. Once the diagram has been developed, the various causes are considered (on the basis of priority) to determine root causes and separate studies. It may be necessary to construct a second-level cause-and-effect diagram as a step in identifying root causes.

QUESTIONS for DISCUSSION

14-1 The All Saints hospital is interested in applying the concept of TQM to bring down its high rate of Cesarean sections (C-sections). Increased patient satisfaction and a decrease in the patient's risk are among the reasons for selecting this problem. A study of the problem revealed the following causes:

Cause	Frequency of occurrence
1. Cephalo-pelvic disproportion	40
2. Infants in breech position	16
3. Fetal distress	14
4. Previous birth by C-section	13
5. Other causes	17

Perform a Pareto analysis using the given data.

14-2 The following data show the average times it has taken a local bank to process loan applications (in hours) for the last 36 applications processed. Develop a histogram using the data. What deductions can you make from the histogram?

10	15	8	40	35	22
62	40	47	22	33	52
11	17	28	18	39	43
31	39	36	12	42	15
43	19	11	62	65	40
13	18	14	21	26	12

14-3 Select a problem from any of the following processes on a university campus, and gather data that would enable you to utilize the tools discussed in this chapter:

▪ Cafeteria
▪ University bookstore
▪ Registration process
▪ Financial aid services
▪ Library

ENDNOTES

1. J. Banks, *Principles of Quality Control*, New York: John Wiley & Sons, 1989.
2. V. K. Omachonu, *Total Quality and Productivity Management in Health Care Organizations*, Norcross, Ga.: Institute of Industrial Engineers, 1991.
3. H. Gitlow, S. Gitlow, A. Oppenheim, and R. Oppenheim, *Tools and Methods for the Improvement of Quality*, Homewood, Ill.: Richard D. Irwin, 1989.
4. M. K. Hart and R. F. Hart, *Quantitative Methods for Quality and Productivity Improvement*, Milwaukee: ASQC Quality Press, 1989.

15

UNDERSTANDING
PROCESS VARIATION

WHAT IS a PROCESS?

The concept of a process was introduced in Chapter 12. When referring to a process, the production or manufacturing process most often comes to mind. In reality, numerous parts of everyday life are, in fact, processes. Filling out an application, visiting a doctor's office, or making a deposit at a bank all are processes. All processes can be observed and improved using statistical methods.

The ultimate goal of studying a process is improvement. In order to improve it, the process must be understood completely, and therefore, it must be documented. In order to document a process, it is necessary to know who is in charge of it, what other process(es) it interacts with, its objectives, and its flow.

The flow of the process is frequently documented graphically using a flowchart. A flowchart is composed of a combination of standard symbols shown in Figure 15-1.

A flowchart provides a general view of a process and helps to explain the roles and interactions between the parts of the process. A flowchart for the process of visiting a doctor's office is provided in Figure 15-2.

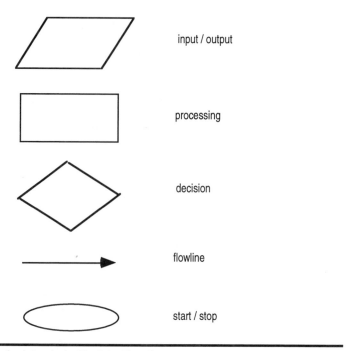

Figure 15-1 Standard Symbols Used in Flowcharts

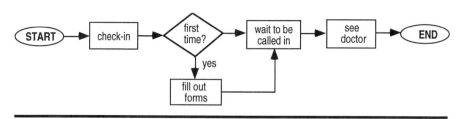

Figure 15-2 Flowchart for the Process of Visiting a Doctor's Office

PROCESS VARIATION

Every process, regardless of how good it is, has variation. Variation is due to either chance causes (common causes) or assignable causes (special causes). Common causes of variation are natural, random, and will always occur; examples include ambient temperature, poor lighting, etc. Special causes, however, are due to some variability in a machine, person, material, environment, or any other component of the process. These are causes that must be detected and controlled in order to keep process variation to a minimum so that the product or service produced will be more constant and uniform.

Shewhart developed the control chart as a statistical tool. Deming uses the control chart as a management tool. Control charts can be used for either purpose, because they point out variations due to special causes and omit variations due to common causes or "noise."

CONTROL CHARTS

A control chart is a run chart with control limits. Control limits are represented as lines above and below the center line. Control limits are computed from the process data and reflect the range of the common causes of variation present in the output of a process. A control chart is a statistical tool based on normal distribution. The normal distribution is represented in the form of a bell-shaped curve. The area under the normal curve is equal to 1 (or 100 percent). The curve is symmetrical about its mean value, and it proves to be a practical interpretation of the standard deviation.[2] The curve encompasses $\pm3\sigma$. The area under a particular interval of the curve represents the probability of obtaining a value within that interval. Control charts have a center line and control limits are set at $\pm3\sigma$.

The control chart, therefore, is effective for detecting the presence of special causes of variation (an unstable or out-of-control condition in a process). It is used to monitor, reduce the variability of, and estimate the parameters of a process. The purpose of a control chart is to differentiate between special and common causes of variation and show the capability of a process. Control charts (and the processes they monitor) are said to be out of control if any point lies outside either of the control limits or if any non-random pattern exists. To detect these non-random patterns, guidelines known as AT&T run rules are used. The control chart is divided into three zones on each side of the center line, each encompassing one sigma, and labeled as follows:

1. Mark with an X any point that falls outside of zone A.

2. If two of the three points in zone A (on the same side of the center line) or beyond are successive, mark the second point. *Note:* The third point can be anywhere on the control chart, including the other side of the center line.

3. If four of the five points in zone B or beyond are successive, mark the fourth point.
4. If there are eight successive points in zone C (on the same side), mark the eighth point.

These rules help reveal points which reflect non-random patterns and thus must be studied.

VARIABLE CONTROL CHARTS

Variable control charts control the mean value of a quality characteristic via an X chart and its variability via an R or S chart. An R chart is used when the sample size is less than or equal to ten. The general terminology used in control charting is as follows:

m = number of samples or subgroups

n = number of observations in each sample

X = value (measurement) of an individual piece

$\bar{X}$ = sample mean = $\dfrac{\Sigma X}{n}$

$\bar{\bar{X}}$ = average of all the sample means = $\dfrac{\Sigma X}{n}$

R = sample range = max X – min X

$\bar{R}$ = average of all the sample ranges = $\dfrac{\Sigma R}{m}$

S = sample standard deviation = $\left[\dfrac{\Sigma(X - X)^2}{n - 1} \right]^{1/2}$

$\bar{S}$ = average of all the sample standard deviations = $\dfrac{\Sigma S}{m}$

UCL = upper control limit

LCL = lower control limit

Examples of the methodology for constructing these charts are provided in the following sections.

$\bar{X}$ and R Charts

The R chart monitors within-sample variability, while the X chart monitors between-sample variability. The X chart is of no use if the R chart is out of control. Thus, the R chart should be plotted before the X chart.

Example

For the purpose of illustration, assume that ten samples of five observations each were taken (a minimum of 25 samples is recommended) of the length of electrical piping in feet. The data are as follows:

$$n = 5, \quad m = 10$$

Sample	X_1	X_2	X_3	X_4	X_5	$\bar{X}$	R
1	10.0	9.7	9.7	9.9	9.7	9.8	0.3
2	9.9	10.1	10.0	9.9	10.3	10.04	0.2
3	9.6	9.9	9.8	9.7	9.6	9.72	0.3
4	9.7	10.2	9.8	9.8	9.8	9.86	0.5
5	9.4	9.6	9.6	9.5	9.8	9.58	0.4
6	10.2	10.1	10.0	10.1	10.2	10.12	0.2
7	10.0	9.8	9.9	10.0	10.3	10.0	0.5
8	9.7	9.6	10.0	9.9	9.7	9.78	0.4
9	9.7	9.9	9.8	10.0	10.0	9.88	0.3
10	10.0	10.2	9.9	10.0	9.8	9.98	0.4

$$\bar{\bar{X}} = 9.88 \qquad \bar{R} = 0.35$$

Calculation of $\bar{X}$ for sample 1:

$$\bar{X} = \frac{\Sigma X_i}{n} = \frac{10.0 + 9.7 + 9.7 + 9.9 + 9.7}{5} = 9.8$$

Calculation of R for sample 1:

$$R = \max X_i - \min X_i = 10.0 - 9.7 = 0.3$$

$$\bar{\bar{X}} = \frac{\text{sum } \bar{X}}{10}$$

$$\bar{R} = \frac{\text{sum } R}{10}$$

The equations for the $\bar{X}$ and R charts based on the sample data are

$\bar{X}$ chart	R chart
UCL $= \bar{\bar{X}} + A_2\bar{R}$	UCL $= \bar{R}D_4$
center line $= \bar{\bar{X}}$	center line $= \bar{R}$
LCL $= \bar{\bar{X}} - A_2\bar{R}$	LCL $= \bar{R}D_3$

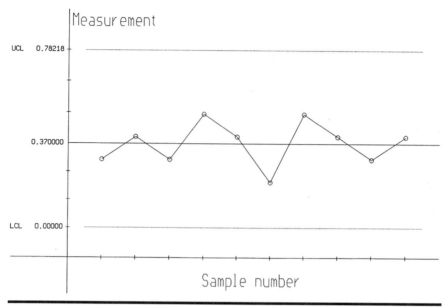

Figure 15-3   R Chart for Length of Electrical Piping

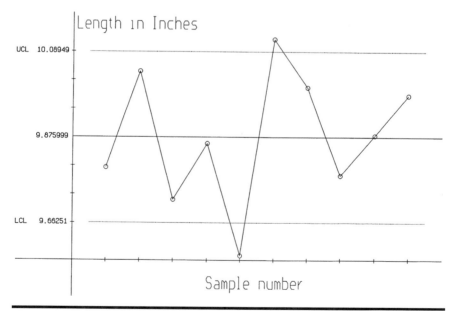

Figure 15-4 $\overline{X}$ Chart for Length of Electrical Piping

The values for A_2, D_3, and D_4 are provided in Appendix A, according to the number of observations, n, in the sample.

R chart values:

$$UCL_R = (0.35)(2.114) = 0.7399$$

$$\text{center line} = 0.35$$

$$LCL_R = (0.35)(0) = 0$$

Being that the R chart is in control, calculation of the X chart follows:

$$UCL_X = 9.88 + (0.577)(0.35) = 10.08$$

$$\text{center line} = 9.88$$

$$LCL_X = 9.88 - (0.577)(0.35) = 9.68$$

Examples of R and $\overline{X}$ charts are provided in Figures 15-3 and 15-4, respectively.

$\overline{X}$ and S Charts

The S chart is used instead of the R chart when the sample size is greater than ten. It is also useful when the sample sizes are unequal.[3]

Example

The following data were taken for the weight (in ounces) of 6-oz. bottles of cough syrup:

Sample	X_1	X_2	X_3	X_4	X_5	X_6	X_7	X_8	X_9	X_{10}	X_{11}	X_{12}	$\overline{X}$	S
1	6.1	6.08	6.1	6.07	6.11	6.12	6.0	6.1	6.2	5.9	5.98	6.02	6.07	0.08
2	6.11	6.06	6.1	6.11	5.96	6.0	6.1	6.12	5.96	6.0	6.01	6.07	6.05	0.06
3	6.12	6.09	6.08	6.04	5.97	6.0	6.11	6.07	5.9	5.92	5.9	6.09	6.02	0.08
4	5.99	6.1	6.07	6.04	5.9	5.93	6.10	6.05	6.01	6.1	5.91	6.04	6.02	0.07
5	5.9	6.0	6.1	6.2	6.13	6.1	6.02	6.06	6.0	6.17	5.97	6.03	6.05	0.08

$$\overline{\overline{X}} = 6.04 \quad \overline{S} = 0.07$$

Calculation of S for the first sample:

$$S^2 = \frac{(6.1 - 6.07)^2 + (6.08 - 6.07)^2 + (6.1 - 6.07)^2 + (6.02 - 6.07)^2}{12 - 1}$$

$$S^2 = \frac{0.068}{11} = 0.006$$

$$S = 0.08$$

The equations for the $\overline{X}$ and S charts based on sample data are

$\overline{X}$ chart	S chart
UCL = $X + A_3\overline{S}$	UCL = $\overline{S}B_4$
center line = X	center line = $\overline{S}$
LCL = $X + A_3\overline{S}$	LCL = $\overline{S}B_3$

Values for A_3, B_3, and B_4 can be found in Appendix A.
    S chart values:

$$\text{UCL}_S = (1.646)(0.07) = 0.12$$

$$\text{center line} = 0.07$$

$$\text{LCL}_S = (0.354)(0.07) = 0.02$$

X chart values:

$$\text{UCL}_X = 6.04 + (0.886)(0.07) = 6.13$$

$$\text{center line} = 6.04$$

$$\text{LCL}_X = 6.04 - (0.886)(0.07) = 6.01$$

Examples of S and $\overline{X}$ charts are provided in Figures 15-5 and 15-6, respectively.

Individuals or X Chart

In addition to the previously mentioned variables control charts, an additional chart is the X chart or individuals chart. It is used when the sample size is equal to one, such as might be the case with very expensive products. The X chart is used in conjunction with a moving range (MR) chart, usually with subgroups of size two. The X chart is less

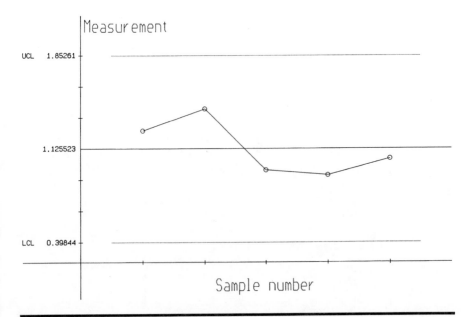

Figure 15-5 S Chart for Diameter of Surgical Tubing

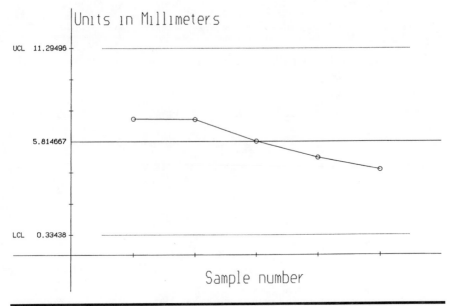

Figure 15-6 X̄ Chart for Diameter of Surgical Tubing

satisfactory than $\overline{X}$ charts, but sometimes must be used. To illustrate the use of X and MR charts, consider the following example.

Example

The local fire department has decided to monitor the closing rate (in seconds) of a local department's four water pump valves. Because only one valve can be observed by one chart, the sample size is one. Therefore, four individuals charts, one for each valve, will be needed. The following are the data for one of the valves:

Sample #	Closing rate (s)	Moving range
1	1.0	
2	1.2	1.1
3	1.1	1.15
4	1.1	1.1
5	1.2	1.15
6	0.9	1.05
7	0.9	0.9
8	1.4	1.15
9	1.4	1.4
10	1.3	1.35
	$\overline{X} = 1.15$	$\overline{MR} = 1.035$

The moving range is calculated by taking an average of two closing rates. The first moving range, 1.1, is an average of 1.0 and 1.2. The second moving range, 1.15, is an average of 1.2 and 1.1, and so on. The control limits for the X chart are calculated using the following equations:

$$\text{UCL}_X = X + 3\left(\frac{MR}{d_2}\right)$$

$$\text{center line} = X$$

$$\text{LCL}_X = X - 3\left(\frac{MR}{d_2}\right)$$

Thus, the control limits for the X chart are

$$UCL_X = 1.15 + 3\left(\frac{1.035}{1.128}\right) = 3.9$$

center line $= 1.15$

$$LCL_X = 1.15 - 3\left(\frac{1.035}{1.128}\right) = -1.6 \rightarrow 0$$

The equations for the MR chart are

$$UCL_{MR} = MR(D_4)$$

center line $= MR$

$$LCL_{MR} = MR\ (D_3)$$

For this example, then, the control limits for the MR chart are

$$UCL_{MR} = (1.035)(3.267) = 3.381$$

center line $= 1.035$

$$LCL_{MR} = (1.035)(0) = 0$$

The values for d_2, D_4, and D_3 are found in Appendix A. The X and MR charts are shown in Figures 15-7 and 15-8, respectively.

ATTRIBUTE CONTROL CHARTS

There are four types of control charts for attributes:

1. p chart (for non-conforming fraction)
2. np chart (for non-conforming number)
3. c chart (for a number of non-conformities)
4. u chart (for a number of non-conformities per unit)

When an item is said to be non-conforming, it means that it is defective and, therefore, unusable. Non-conformities, on the other hand, refer to defects that an item may have but which do not necessarily make the item unusable or defective. *Note:* Attribute charts are not based on normal distribution, as are variable control charts. However, the normal

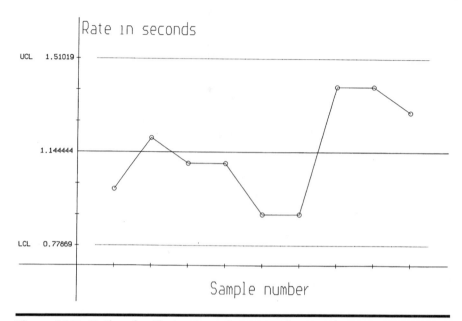

Figure 15-7   X Chart for Closing Rate of Fire Department Valves

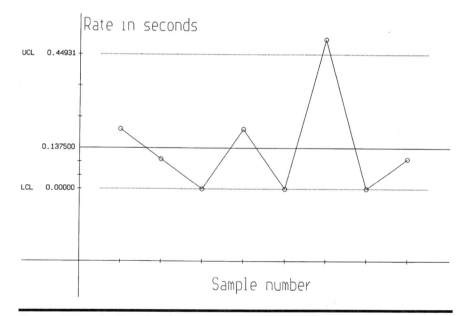

Figure 15-8   MR Chart for Closing Rate of Fire Department Valves

Table 15-1 Control Limits for Attribute Control Charts

p chart $\text{UCL} = p + 3\left[\dfrac{p(1-p)}{n}\right]^{1/2}$ $\text{UCL} = \bar{p} + 3\left[\dfrac{\bar{p}(1-\bar{p})}{n}\right]^{1/2}$

center line $= p$ center line $= \bar{p}$

$\text{LCL} = p - 3\left[\dfrac{p(1-p)}{n}\right]^{1/2}$ $\text{LCL} = \bar{p} - 3\left[\dfrac{\bar{p}(1-\bar{p})}{n}\right]^{1/2}$

np chart $\text{UCL} = np + 3[np(1-p)]^{1/2}$ $\text{UCL} = n\bar{p} + 3[n\bar{p}(1-\bar{p})]^{1/2}$

center line $= np$ center line $= n\bar{p}$

$\text{LCL} = np - 3[np(1-p)]^{1/2}$ $\text{LCL} = n\bar{p} - 3[n\bar{p}(1-\bar{p})]^{1/2}$

c chart $\text{UCL} = c + 3[c]^{1/2}$ $\text{UCL} = \bar{c} + 3[\bar{c}]^{1/2}$

center line $= c$ center line $= \bar{c}$

$\text{LCL} = c - 3[c]^{1/2}$ $\text{LCL} = \bar{c} - 3[\bar{c}]^{1/2}$

u chart $\text{UCL} = u + 3\left[\dfrac{u}{n}\right]^{1/2}$ $\text{UCL} = \bar{u} + 3\left[\dfrac{\bar{u}}{n}\right]^{1/2}$

center line $= u$ center line $= \bar{u}$

$\text{LCL} = u - 3\left[\dfrac{u}{n}\right]^{1/2}$ $\text{LCL} = \bar{u} - 3\left[\dfrac{\bar{u}}{n}\right]^{1/2}$

distribution can be used as an acceptable practical basis for attribute charts.[2]

The equations for attribute control charts which will be used in the examples that follow are listed in Table 15-1.

p Chart

The p chart is based on the binomial distribution, as is the np chart. p represents the non-conforming faction This parameter is estimated in a sample by

$$p = \frac{D}{n}$$

where D = the number of non-conforming units and n = the number of units in the sample. Consider the following example.

The following data correspond to the number of tennis rackets which were strung improperly in the production process. Each sample had 50 observations (n = 50).

Sample #	# non-conforming	Sample fraction non-conforming $\hat{p}$
1	12	0.24
2	15	0.3
3	6	0.12
4	3	0.06
5	10	0.2
6	9	0.18
7	6	0.12
8	11	0.22
9	20	0.4
10	17	0.34
	Σ = 109	$\bar{p}$ = 0.218

$$UCL_p = 0.218 + \left[\frac{(0.218)(0.782)}{50}\right]^{1/2} = 0.393$$

center line = 0.218

$$LCL_p = 0.218 - 3\left[\frac{(0.218)(0.782)}{50}\right]^{1/2} = 0.043$$

Note: Should the LCL equal a negative number, then it is set at 0, because it is not possible to have a negative non-conforming fraction.

Refer to Figure 15-9 for the corresponding p chart.

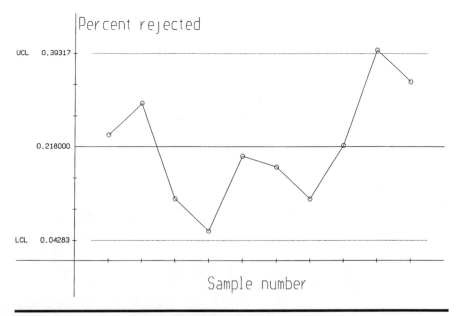

Figure 15-9 p Chart for Tennis Rackets Strung Improperly

np Chart

The np chart is at times easier to understand than the p chart, because it refers to a number, rather than a percentage. An np chart is calculated in the same way as a p chart, except that the control limits are multiplied by the number of observations, n. Using the same data as that for the preceding p chart,

$$UCL_{np} = (50)(0.218) + 3[(50)(0.218)(0.782)]^{1/2} = 21.8$$

$$\text{center line} = (50)(0.218) = 10.9$$

$$LCL_{np} = (50)(0.218) - 3[(50)(0.218)(0.782)]^{1/2} = 2.14$$

Note: In order to use an np chart, the number of observations must be constant for every sample. This is not the case for the p chart.

c Chart

The following table presents the data for the number of scratches on windows being produced for new housing projects. Each sample had 20 observations.

Sample #	# of non-conformities (c)
1	10
2	5
3	16
4	12
5	3
6	15
7	17
8	9
9	9
10	7
	$\Sigma = 103$

$$c = \frac{103}{10} = 10.3$$

$$\text{UCL}_c = 10.3 + 3[10.3]^{1/2} = 19.93$$

$$\text{center line} = 10.3$$

$$\text{LCL}_c = 10.3 - 3[10.3]^{1/2} = 0.67$$

Note: The sample size must be constant for a c chart.

u Chart

Using the same data as for the c chart, the u chart is calculated as follows. The sample size is ten for every sample taken. This need not be constant for the use of a u chart.

Sample #	# of non-conformities (c)	# of non-conformities per unit (u)
1	10	1
2	5	0.5
3	16	1.6
4	12	1.2
5	3	0.3
6	15	1.5
7	17	1.7
8	9	0.9
9	9	0.9
10	7	0.7
	$\Sigma = 103$	$\Sigma = 10.3$

$$c = \frac{10.3}{10} = 1.03$$

$$\text{UCL}_u = 1.03 + 3\left[\frac{1.03}{10}\right]^{1/2} = 1.99$$

$$\text{center line} = 1.03$$

$$\text{LCL}_u = 1.03 - 3\left[\frac{1.03}{10}\right]^{1/2} = 0.067$$

The conditions for choosing when to use each type of control chart are summarized in Figure 15-10.

QUESTIONS for DISCUSSION

15-1 What is a process? Select a process that you know and identify its key inputs and outputs.

15-2 Explain the following terms:
- Process variation
- Process capability
- Control limits
- Mean
- Variance
- Standard deviation

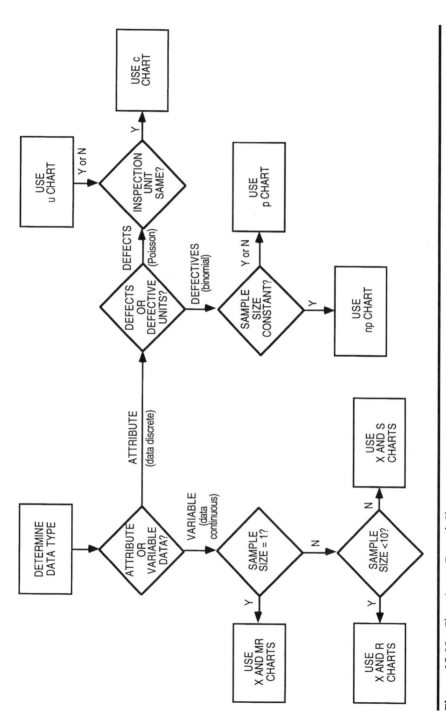

Figure 15-10 Choosing a Control Chart

APPENDIX A
CONTROL CHART CONSTANTS

Number of observations in subgroup, n	A_2	A_3	B_3	B_4	c_4	d_2	d_3	D_3	D_4	E_2
2	1.880	2.659	0.000	3.267	0.7979	1.128	0.853	0.000	3.267	2.660
3	1.023	1.954	0.000	2.568	0.8862	1.693	0.888	0.000	2.574	1.772
4	0.729	1.628	0.000	2.266	0.9213	2.059	0.880	0.000	2.282	1.457
5	0.577	1.427	0.000	2.089	0.9400	2.326	0.864	0.000	2.114	1.290
6	0.483	1.287	0.030	1.970	0.9515	2.534	0.848	0.000	2.004	1.184
7	0.419	1.182	0.118	1.882	0.9594	2.704	0.833	0.076	1.924	1.109
8	0.373	1.099	0.185	1.815	0.9650	2.847	0.820	0.136	1.864	1.054
9	0.337	1.032	0.239	1.761	0.9693	2.970	0.808	0.184	1.816	1.010
10	0.308	0.975	0.284	1.716	0.9727	3.078	0.797	0.223	1.777	0.975
11	0.285	0.927	0.321	1.679	0.9754	3.173	0.787	0.256	1.744	
12	0.266	0.886	0.354	1.646	0.9776	3.258	0.778	0.283	1.717	
13	0.249	0.850	0.382	1.618	0.9794	3.336	0.770	0.307	1.693	
14	0.235	0.817	0.406	1.594	0.9810	3.407	0.762	0.328	1.672	
15	0.223	0.789	0.428	1.572	0.9823	3.472	0.755	0.347	1.653	
16	0.212	0.763	0.448	1.552	0.9835	3.532	0.749	0.363	1.637	
17	0.203	0.739	0.466	1.534	0.9845	3.588	0.743	0.378	1.622	
18	0.194	0.718	0.482	1.518	0.9854	3.640	0.738	0.391	1.608	
19	0.187	0.698	0.497	1.503	0.9862	3.689	0.733	0.403	1.597	
20	0.180	0.680	0.510	1.490	0.9869	3.735	0.729	0.415	1.585	
21	0.173	0.663	0.523	1.477	0.9876	3.778	0.724	0.425	1.575	
22	0.167	0.647	0.534	1.466	0.9882	3.819	0.720	0.434	1.566	
23	0.162	0.633	0.545	1.455	0.9887	3.858	0.716	0.443	1.557	
24	0.157	0.619	0.555	1.445	0.9892	3.895	0.712	0.451	1.548	
25	0.153	0.606	0.565	1.435	0.9896	3.931	0.709	0.459	1.541	
More than 25	$3/\sqrt{n}$		$1-3/\sqrt{2n}$	$1+3/\sqrt{2n}$						

Copyright ASTM. Reprinted with permission, from *ASTM Manual on the Presentation of Data and Control Chart Analysis* (1976, pgs. 134–136).

ENDNOTES

1. H. Gitlow, S. Gitlow, A. Oppenheim, and R. Oppenheim, *Tools and Methods for the Improvement of Quality,* Homewood, Ill.: Richard D. Irwin, 1989.
2. M. Owen, *PC and Continuous Improvement,* United Kingdom: IFS Publications, 1989.
3. J. Banks, *Principles of Quality Control,* New York: John Wiley & Sons, 1989.
4. M. K. Hart and R. F. Hart, *Quantitative Methods for Quality and Productivity Improvement,* Milwaukee: ASQC Quality Press, 1989.
5. V. K. Omachonu, *Total Quality and Productivity Management in Health Care Organizations,* Norcross, Ga.: Institute of Industrial Engineers, 1991.

III

CRITERIA for
QUALITY PROGRAMS

In Part III, the commonly accepted standards for measuring the effectiveness of an organization's quality program are presented. ISO 9000 is the most widespread standard in Europe. It is the system adopted by the International Organization for Standardization (ISO) and is accepted by all member nations of the European Community. For this reason, it is almost mandatory for firms planning to do business in Europe.

The criteria contained in the Baldrige Award are much more comprehensive and are widely used as the benchmark in the United States. Firms planning to implement TQM should consider adopting these criteria as an objective to be achieved.

The concept of reengineering is examined in Chapter 18.

16

ISO 9000:
UNIVERSAL STANDARDS
of QUALITY

Companies can comply with Europe's standards—or stay home

Business Week

"Simply put, ISO 9000 has come to be the price of admission for doing business in Europe," says Robert Caine, president of the American Society for Quality Control (ASQC). "Ask any business person who has given up trying to gain entry into the European market what stopped him, and he's likely to answer in code: ISO 9000," concludes Kymberly Hockman of Du Pont's Quality Management and Technology Center. These are among the many experts who are urging U.S. firms to take the ISO Series standards seriously.

Even if a firm does not do business in Europe or does not plan to do so, it should not ignore this accelerating movement to international standards. As will be discussed, the movement is expanding into other areas of the world and into many areas of the U.S. public and private sectors as well.

ISO 9000 is a set of five worldwide standards that establish requirements for the management of quality. Unlike *product* standards, these standards are for *quality management systems*. They are being used by the twelve-nation European Economic Community to provide a universal framework for quality assurance—primarily through a system of internal and external audits. The purpose is to ensure that a certified company has a quality system in place that will enable it to meet its published quality standards. The ISO standards are generic in that they apply to all functions and all industries, from banking to chemical manufacturing. They have been described as the "one size fits all" standards.

ISO around the WORLD

The European Community (EC) consists of twelve member nations: Belgium, Denmark, France, Germany, Greece, Ireland, Italy, Luxembourg, the Netherlands, Portugal, Spain, and the United Kingdom. The goal of the EC is to create a single internal market, free of all barriers to trade. For products and services to be traded freely, there must be assurance that those product meet certain standards, whether they are produced in one of the EC nations or in a non-EC nation, such as the United States.[1] The EC is using the standards to provide a universal framework for quality assurance and to ensure the quality of goods and services across borders.

The International Organization for Standardization (ISO) is the specialized international agency for standardization and at present comprises the national standards bodies of 91 countries. The American National Standards Institute (ANSI) is the member body representing the United States. ISO is made up of approximately 180 technical committees. Each technical committee is responsible for one of many areas of specialization, ranging from asbestos to zinc. The purpose of ISO is to promote the development of standardization and related world activities in order to facilitate the international exchange of goods and services and to develop cooperation in intellectual, scientific, technological, and economic activities. The results of ISO technical work are published as international standards and the ISO 9000 Series is a result of this process.

In 1987 (the same year the ISO 9000 Series was published), the United States adopted the ISO 9000 Series verbatim as the ANSI/ASQC Q-90 Series. Thus, the use of either of these series is equivalent to the use of the other.[2] The ISO standards are being adopted by a varying number of companies in over 50 countries around the world that have endorsed them. Many people believe that within five years registration will be necessary to stay in business.[3]

By 1992 more than 20,000 facilities in Britain had adopted the standards and became certified.[4] Over 20,000 companies from other EC countries have registered, compared to about 620 in the U.S. The Japanese not only have adopted the standards, but also have mounted a major national effort to get their companies registered.[5]

The EC adopted ISO 9000 in 1989 to integrate the various technical norms and specifications of its member states. By 1991, ISO compliance became part of hundreds of product safety laws all over Europe, regulating everything from medical devices to telecommunications gear. Such products accounted for only about 15 percent of EC trade at that time, but the list of products is growing. Entire industries are encouraging the adoption of the standards.

One example of the impact is reflected in the requirements of Siemens, the huge German electronics firm. The company requires ISO compliance in 50 percent of its contracts and is pressing all other suppliers to conform. A major justification for this action is that it eliminates the need to test parts, which saves time and money and establishes common requirements for all markets.

Even for companies whose products are unregulated, ISO standards are becoming a de facto market requirement for doing business with other EC companies. If two suppliers are competing for a contract or an order, the one that has registered its quality systems under ISO 9000 has a clear edge.

The impact of these standards is reflected by the widespread distribution of the ISO 9000 Series, which has become the best-seller in the history of the ISO, under whose auspices they were developed. ISO 9000 even outsold the universal and long-standing international weights and measurement standards. However, it is worth repeating that ISO 9000 is not standards for products, but standards for operation of a *quality management system.*

ISO 9000 in the UNITED STATES

U.S. companies have been slow to adopt these international standards despite the fact that 30 percent of the country's exports go to Europe. Moreover, to the extent that the standards are adopted elsewhere in the world, additional exports will be affected as well. Additional markets both within and outside the United States may be closed to those firms that ignore the requirement or fail to be certified. Du Pont, now a leader in adopting the standards, only began its ISO drive in 1989 after losing a large European order for polyester film to an ISO-certified British firm.

Some people perceive ISO 9000 as a barrier to competition and even a plot to keep U.S. firms out of Europe. This view, of course, is not the case, but a barrier can exist unless the standards are clearly understood.

Additional evidence of growing acceptance lies in the fact that the standards are being integrated into the requirements for manufacturers that make products under contract for several U.S. government agencies, including NASA, the Department of Defense, the Federal Aviation Administration, and the Food and Drug Administration.[6] To date, ISO 9000 registration is required of suppliers to the governments of Canada, Australia, and the U.K.

Du Pont, Eastman Kodak, and other U.S. pioneers adopted ISO 9000 in the late 1980s to ensure that they were not locked out of European markets. They then found that the standards also helped to improve their quality. Now, Baldrige winners such as Motorola, Xerox, IBM, and others are making suppliers adopt ISO. As the movement catches on and as suppliers to suppliers are required to come on board, there may be a geometric leverage effect in the number of companies adopting the standards. This effect may give additional meaning to the often-repeated description of the market as *global* in dimension.

Despite the weight of the evidence that suggests the need to adopt ISO 9000, it appears that many U.S. firms have not done so, nor do they plan to do so. One survey of 254 mid-sized manufacturing firms conducted by the Chicago accounting firm of Grant Thornton found that only 8 percent planned to become certified by the end of 1992 and 48 percent of the senior executives never even heard of the ISO 9000 standards.[7]

The good news is that for those firms planning to become ISO 9000 certified, the process is not all that difficult, especially if the company already has a quality effort underway. Indeed, those companies using total quality management (TQM) are more than half way there. For Baldrige winners, certification would be a relatively simple process.

What is the impact of ISO 9000 for service industries and for those manufacturing firms whose products fall outside the *regulated* product areas? The answer is provided by ASQC:[8]

> Outside of regulated product areas, the importance of ISO 9000 registration as a competitive market tool varies from sector to sector. For instance, in some sectors, European companies may require suppliers to attest that they have an approved quality system in place as a condition for purchase. This could be specified in any business contract. ISO 9000 registration may also serve as a

means of differentiating "classes" of suppliers, particularly in high-tech areas, where high product reliability is crucial. In other words, if two suppliers are competing for the same contract, the one with ISO 9000 registration may have a competitive edge with some buyers. Sector and product areas where purchasers are more likely to generate pressure for ISO 9000 registration include aerospace, autos, electronic components, measuring and testing instruments, and so on. ISO 9000 registration may also be a competitive factor in product areas where safety or liability are concerns.

Some American manufacturers have criticized the EC's adoption of ISO 9000, suggesting that the standards are inferior to those used in the United States. Moreover, it is suggested that requiring U.S. companies to conform to the standards will force them to incur larger production costs.[9]

The counter arguments are that the standards will eliminate the hodgepodge of standards that now exist around the world, and production costs will be more than offset by other savings and the increase in productivity and quality.

Criticisms and ignorance of ISO 9000 notwithstanding, there is evidence of a growing acceptance of the standards among U.S. firms. One source reports an increase in registration of 500 percent between 1992 and 1993. Of course, this increase is computed on a somewhat smaller 1992 base.[10] It is interesting to note that the Japanese experience is similar to that in the United States. Initial resistance was largely overcome by pressure to conform to the requirements of the international marketplace.[11]

Involvement of professional and trade associations appears to be growing as firms within a particular industry band together to research how best to meet ISO requirements. The chemical industry has been a leader in this movement. Professional engineers, public utilities, software vendors, and manufacturers of information technology are among the groups with organized efforts.[12] Some have formed a network of support groups.[13]

The ISO 9000 ANSI/ASQC Q-90 SERIES STANDARDS[14]

Unlike the Baldrige, the ISO 9000 Series and its clone, the ANSI/ASQC Q-90 Series, are not awards programs. They do not require the use of any state-of-the-art system, nor do they require any prescribed method

Table 16-1 Summary of ISO 9000 Standards

Standard	Content	Application
ISO 9000	Provides definitions and concepts Explains how to select other standards for a given business	All industries including software development
ISO 9001	Quality assurance in design, development, production, installation, and servicing	Engineering and construction firms, manufacturers that design, develop, install, and service products
ISO 9002	Quality assurance in production and installation	Companies in the chemical process industries that are not involved in product design or after-sales service
ISO 9003	Quality assurance in test and inspection	Small shops, divisions within a firm, equipment distributors that inspect and test supplied products
ISO 9004	Quality management and quality system elements	All industries

of process control. They are generic and apply to all industries.[15] As a set of requirements for quality systems, these series provide a common measuring stick for gauging quality systems. Leaving the determination of quality levels to the customer–supplier interaction, the series fill the need for a customer's guarantee that a supplier will, within defined limits, be able to deliver products and services as promised.[16] This flexibility and lack of constraining requirements mean that there is no one right way to do ISO 9000. Industries are free to find their own way and perceive this as an opportunity rather than an additional constraint. This freedom can serve as a source of both frustration as well as liberation.

The ISO 9000 Series is so named because it consists of *five* sets of standards, numbered sequentially from 9000. A brief summary of each standards is provided in Table 16-1.

ISO 9001 ensures conformance to requirements during design, development, production, installation, and service. The quality systems requirements[17] are

Management responsibility	Inspection, measuring, and test equipment
Quality system	Inspection and test status
Contract review	Control of non-conforming product

Design control

Document control

Purchasing

Purchaser-supplied product

Product identification
and traceability

Process control

Inspection and testing

Corrective action

Handling, storage, packaging, and delivery

Quality records

Internal quality audits

Training

Servicing

Statistical techniques

Documentation

Many firms hesitate to comply with the standards due to the onerous task of *documentation*. The advice given to some firms is to "document what you do and do what you document." Nevertheless, documentation actions to comply with the standards are necessary. There are three major tasks:

1. Write the quality manual according to ISO guidelines
2. Document all relevant procedures
3. Write all relevant work instructions

An excerpt from a standard on documentation is as follows.

4.5 Document Control

4.5.1 Document Approval and Issue

The supplier shall establish and maintain procedures to control all documents and data that relate to the requirements of this Standard. These documents shall be reviewed and approved for adequacy by authorized personnel prior to issue. This control shall ensure that: a) the pertinent issues of appropriate documents are available at all locations where operations essential to the effective functioning of the quality system are performed; b) obsolete documents are promptly removed from all points of issue or use.

4.5.2 Document Changes/Modifications

Changes to documents shall be reviewed and approved by the same functions/organizations that performed the original review and approval unless specifically designated otherwise. The designated organizations shall have access to pertinent background information upon which to base their review and approval.

Where practicable, the nature of the change shall be identified in the document or the appropriate attachments.

A master list of equivalent document control procedures shall be established to identify the current revision of documents in order to preclude the use of non-applicable documents.

Documents shall be re-issued after a practical number of changes have been made.

Management Responsibility

The commitment and involvement of top management are requirements for the success of any significant cultural or operational change. So it is with both the Baldrige and ISO 9000. The concern of ISO with management responsibility is reflected in the following series excerpts:[18]

Quality policy. The supplier's management shall define and document its policy and objectives for, and commitment to, quality. The supplier shall ensure that this policy is understood, implemented, and maintained at all levels in the organization.

Management review. The quality system adopted to satisfy the requirement of the standard shall be reviewed at appropriate intervals by the supplier's management to ensure its continuing suitability and effectiveness. Records of such reviews shall be maintained.

Internal quality audits. The supplier shall carry out a comprehensive system of planned and documented internal quality audits to verify whether quality activities comply with planned arrangements and to determine the effectiveness of the quality system. Audits shall be scheduled on the basis of the status and importance of the activity. The audits and follow up actions shall be carried out in accordance with documented procedures.

The results of the audits shall be documented and brought to the attention of the personnel having responsibility in the area audited. The management personnel responsible for the area shall take timely corrective action on the deficiencies found by the audit.

Corrective action. The supplier shall establish, document and maintain procedures for:

■ Investigating the cause of nonconforming product and the corrective action needed to prevent recurrence

■ Analyzing all processes, work operations, concessions, quality records, service reports, and customer complaints to detect and eliminate potential causes of nonconforming product

■ Initiating preventive actions to deal with problems to a level corresponding to the risks encountered

■ Applying controls to ensure that corrective actions are taken and that they are effective

■ Implementing and recording changes in procedures resulting from corrective action

Functional Standards

ISO 9000 standards also require documentation and follow-up performance for all functions affecting quality. Functional requirements are illustrated by the following examples:[19]

■ **Design.** Sets a planned approach for meeting product or service specifications

■ **Process Control.** Provides concise instructions for manufacturing or service functions

■ **Purchasing.** Details methods for approving suppliers and placing orders

■ **Service.** Detailed instructions for carrying out after-sales service

■ **Inspection and testing.** Compels workers and managers to verify all production steps

■ **Training.** Specifies methods to identify training needs and keeping records

BENEFITS of ISO 9000 CERTIFICATION

The benefits to the organization gained by improving quality in products and services were outlined in Chapter 1. To repeat:

1. Greater customer loyalty
2. Improvements in market share
3. Higher stock prices
4. Reduced service calls
5. Higher prices
6. Greater productivity and cost reduction

These same benefits would be achieved by ISO 9000 certification to the extent that actions leading to certification result in a quality management system. Moreover, certification provides the additional benefit of acceptance by EC customers and others whose criteria of acceptance include ISO 9000 certification.

Experience tends to confirm that companies do achieve these benefits. Consider the following examples:

▪ A British government survey revealed that 89 percent of ISO 9000 registered companies reported greater operational efficiency: 48 percent reported increased profitability, 76 percent reported improvements in marketing, and 26 percent reported increased export sales.[20]

▪ The British Standards Institution, a leading British Registrar, estimates that registered firms reduce operating costs by 10 percent on average.[21]

▪ Du Pont attributes the following results to the adoption of ISO standards in their plants:

   ▪ On-time delivery at one plant increased to 90 percent from 70 percent

   ▪ Cycle time at one plant went from 15 days to 1 1/2 days

   ▪ First-pass yield at one plant went from 72 percent to 92 percent

   ▪ Test procedures were reduced from 3000 to 1100

▪ A number of U.S. firms have reported benefits ranging from increased sales to improved communications.[22]

GETTING CERTIFIED: The THIRD-PARTY AUDIT

Many managers perceive the thought of an audit of any kind as a necessary bureaucratic action that has a very low priority. This negative perception may increase when it is learned that preparation for ISO 9000 certification may take from six to twelve months and that the failure rate the first time around can be as high as two out of three. Nevertheless, a third-party audit is a prerequisite to certification. Speaking of certification, Deming noted, "You don't have to do this—survival is not compulsory!"

The traditional two-party quality audit system relies on the buyer–seller relationship, where the buyer (customer) "audits" the supplier. This puts a burden on both parties. Imagine a supplier with a hundred or more customers, each with their own specific requirements. From a customer's point of view, it would be beneficial if all suppliers could be judged by a single set of criteria.

The third-party audit places great importance on quality systems, a critical factor in the EC. The independent third-party *registrar* certifies that the quality system meets the requirements of ISO 9000.

What is the rationale for a third-party audit? Financial results are measured by financial statements, while product and service outputs are measured by quality. If the impartial third-party audit is required for financial systems, why not a similar check on quality systems? This is particularly important in helping to guarantee quality across international borders.

DOCUMENTATION

There are three basic steps to the registration process:

1. Appraisal of the organization's quality manual
2. Evaluation of conformance to documented procedures
3. Presentation of findings, with recommendations for corrective action

A great deal of *documentation* is required. The justification is reflected in the management axiom, "if you haven't written it out, you haven't thought it out." Moreover, as people come and go, change jobs, and forget a procedure, documentation ensures that a record is maintained for continuity. The simple rule is that if all personnel involved in a given system or procedure were replaced, the new people could continue making the product at the same quality level.

The amount of documentation depends on the nature and complexity of the business. A hierarchical approach involving three levels is generally acceptable:

■ **Level 1:** An overview type of quality manual consisting of policies that meet the requirements of the ISO standard for which certification is sought.

- **Level 2:** Functional or departmental operating procedures in terms of "who does what."
- **Level 3:** Work instructions that explain how each task is to be accomplished.

The criteria for approval are simple: "Can you say what you do and do what you say?" Questions such as the following may be asked: Is the process control system adequate for your needs? Is it understood by those who run the process? Are they properly trained to operate the process? Is the documentation up to date? Do you have an internal audit system that regularly assesses whether the control system is functioning as it should be?

POST-CERTIFICATION

The third-party audit and subsequent certification, if achieved, should be viewed as a means, not and end to be achieved. The importance of preparation for certification lies not so much in the certification itself, but in the quality system that results from the effort leading to it.

The customer is the ultimate beneficiary of the quality system, and any effort to obtain ISO 9000 certification without customer communication can be a waste of time and a compromise of any system that may result.

Certification is a beginning, not an end. Continuous evaluation, feedback, and fine-tuning are suggested. Who will perform this internal and continuing "audit" following certification? The responsibility, of course, is top management's. The role of the internal auditor, if any, is not clear. Should the role include getting ready for certification or maintaining post-certification requirements, or both?[23] The role is not clearly assigned and may represent an opportunity for internal auditors.

CHOOSING an ACCREDITED REGISTRATION SERVICE

Quality managers who decide to implement an ISO 9000 system are confronted by two related issues: how best to implement the new system and how to ensure that certification will be recognized by customers. This latter issue will normally be settled if certification is recognized by legitimate accreditation bodies.

U.S. firms located in Europe normally utilize one of the many accrediting bodies in those countries. Many are government sanctioned, such as Raad voor de Certificatie (RvC) in the Netherlands and the National Accreditation Council for Certification Bodies (NACCB) in the United Kingdom. IBM's Application Business Systems Division was the first American-based firm to be certified in all of its business lines. Certification was gained after an audit by Bureau Veritas Quality International.[24]

No single firmly established registrar-accredited authority is recognized in the United States, and confusion exists as to which auditors are accredited by whom. Two non-governmental groups—the Registrar Accreditation Board (RAB) (an offshoot of the ASQC) and ANSI—have carried out a joint effort to develop accreditation requirements for ISO 9000 auditing companies operating in the United States.[25] The creation of the ANSI/RAB accreditation program is the nearest source of credible U.S.-based registrars.

A number of criteria should affect the decision on the choice of a registrar, including their knowledge in the specific industry and in the auditing of quality systems, how many similar firms they have registered, their turnaround time for audit results, their re-audit schedule should complement the business cycle of the firm, and, most important, they should be accredited.

As a general rule, it is probably not wise to shop around for the lowest price, because the cost of an audit is small compared to the overall cost of the registration effort.

ISO 9000 and SERVICES

The standards apply not only to the manufacturing process, but to after-sale service and to service departments such as design within the manufacturing firm as well. Additionally, the standards also translate to the service sector. They specifically address quality systems for service as well as production. Indeed, ISO 9000-2, a separate guideline, was issued to explain ISO criteria in terms of selected service industries.

In the United Kingdom, standards are being used by educational institutions, banks, legal and architectural firms, and even trash collectors. At London's Heathrow Airport, British Airways PLC adopted ISO standards to reduce complaints of lost cargo and damaged goods. In the United States, a growing number of transportation companies will not transport hazardous material unless the shipper is ISO certified.

There is some evidence that the ISO 9000 Series is receiving more interest from service organizations in the United States than in Europe. Service firms in consulting, purchasing, and materials management are expressing interest. It is believed by some that the greater interest by U.S. service firms is based on strategic considerations as ISO 9000 is perceived as a "market differentiator."[26]

The COST of CERTIFICATION

A frequently asked question is: "How much does certification cost?" This is a legitimate concern, although the question may be accompanied by another one: "What is the payoff?"

There is no set answer to how much it costs and how long it takes. Each company is different. The answer depends on such factors as company size, product lines, how far along the company's existing systems are on the quality continuum, whether consultants are used, and the implementation strategy adopted. It can cost a small company $2,000 to $25,000 in consulting fees for advice on developing a quality system.[27] Employee time in creating the system is additional and can be the largest cost.

The major determinant is the firm's starting position. If the company has just won a Baldrige Award, registration of a plant or business might take just a few days. However, if the system must be created from the ground up, it can take a year and cost $100,000 or more.[28]

ISO 9000 vs. the BALDRIGE AWARD

The ASQC reports that one of the most frequently asked questions regarding the ISO Series is: "Aren't the Baldrige Award, the Deming Prize, etc. equivalent or better 'standards' than the ISO Series?" The answer, replies ASQC, is quite simple: "You can't hope to meet the expectations of any of these programs if you aren't already implementing the ISO 9000 (ANSI/ASQC Q-90) standards in your company. These standards provide the foundation on which you can build your quality management and quality assurance systems so you may ultimately achieve a high level of success. Moreover, the ISO 9000 Series is the only system accepted internationally."

The Baldrige is a much more comprehensive program than ISO 9000. It is truly a TQM system, whereas the ISO Series is much more limited in scope. It is a basic standard, a minimal requirement, and can be worth about 200 to 300 points in the Baldrige program. For example, it does not address the human resource dimension, as does the Baldrige. On the other hand, a company implementing the Baldrige criteria is in a much better position to implement the ISO standard.

The Baldrige criteria are much more specific. The guidelines spell out what is expected in detailed language. In contrast, the ISO Series is designed to be inclusive, not exclusive. It does not mandate that one approach be used over another. As long as you can say what you do and do what you say, you can get your system registered. This generic nature of the standards can be a source of frustration as well as liberation.

For those companies whose quality systems are on the low end of a TQM continuum, ISO may be a starting place on the road to eventually achieving a TQM system. Certification also has the advantage of putting the organization on a level playing field with the competition worldwide.

IMPLEMENTING the SYSTEM

Although the series provides guidance on the required attributes of the quality system, the standards do not spell out the means of implementation. Once a decision is made to adopt the standards and seek certification, the following major steps will facilitate successful change:

- Recognize the need for change and get the commitment of top management.
- Incorporate quality in the strategic plan as the linchpin of differentiation.
- Formulate and adopt a holistic quality policy statement adapted to ISO requirements. Get support and commitment from all managers.
- Determine the scope of the business to be certified. Will it be a particular process, related facilities, a geographical site, or the whole company?
- Determine the status of the current quality system through an internal audit. Define the *gap* between where you are and what it will take to close the gap.
- Estimate the cost in time and money and implement the plan by organizing the necessary action steps.

QUESTIONS for DISCUSSION

16-1 Why is it important for U.S. firms to comply with ISO 9000?

16-2 Compare the standards of ISO 9000 with those of the Baldrige Award.

16-3 Does ISO 9000 contain product standards or standards for operation of a quality management system? Explain the difference.

16-4 Answer the criticisms that meeting ISO standards will add to production costs.

16-5 What are the five sets of standards? Summarize each.

16-6 What are the benefits of ISO 9000 certification?

ENDNOTES

1. Gary Spizizen, "The ISO 9000 Standards: Creating a Level Playing Field for International Quality," *National Productivity Review,* Summer 1992, p. 332. This is an excellent summary of the provisions of ISO 9000.
2. The ANSI/ASQC Series is available from ASQC headquarters through the customer service department (Tel. 800-248-1946). The ISO 9000 Series is available from ANSI (Tel. 212-642-4900). Keep in mind that the ANSI/ASQC Q-90 Series is *identical* to the ISO 9000 Series.
3. Suzan L. Jackson, "What You Should Know about ISO 9000," *Training,* May 1992, p. 48. This is a good primer on ISO standards. ISO 9000 was adopted by the EC in 1990 as a global standard of quality. Its stringent requirements ensure that products manufactured along ISO 9000 are world class. See Jack Cella, "ISO 9000 Is the Key to International Business," *Journal of Commerce and Commercial,* Jan. 25, 1993, p. 88. Even China is moving toward adoption of the standards according to Ed Haderer, "Setting Tough Standards," *The China Business Review,* Jan.–Feb., 1993, p. 34. The Shanghai-Foxboro Company Ltd., an affiliate of the U.S.-based Foxboro Company, became the first company in China to attain ISO 9000 certification.
4. "Want EC Business? You Have Two Choices," *Business Week,* Oct. 19, 1992, p. 58. In the U.K., where the standards have become most widely embraced, over 80 percent of large employers with payrolls over 1000 are registered. See Kymberly K. Hockman and David A. Erdman, "Gearing Up for ISO 9000 Registration, *Chemical Engineering,* April 1993, p. 128.
5. Donald W. Marquardt, "ISO 9000: A Universal Standard of Quality," *Management Review,* Jan. 1992, p. 50.

6. "U.S. Firms Lag in Meeting Global Quality Standards," *Marketing News*, Feb. 15, 1993.

7. Jeffrey A. Tannenbaum, "Small Companies Are Finding it Pays to Think Global; Firms Win New Business by Adopting International Quality Standards," *Wall Street Journal*, Nov. 19, 1992, Section B, p. 2. See also *Marketing News*, Feb. 15, 1993, p. 1.

8. American Society for Quality Control, "ISO 9000," a brochure prepared by the Standards Development Department of ASQC, P.O. Box 3005, Milwaukee, WI 53201 (Tel. 414-272-8575).

9. Milton G. Allimadi, "New Quality Standards Draw Fire from US Group," *Journal of Commerce and Commercial*, Jan. 4, 1993, p. 4.

10. Mark Morrow, "International Agreements Increase Clout of ISO 9000," *Chemical Week*, April 7, 1993, p. 32. This article attributes the success of ISO 9000 to its brevity (20 pages) and its simplicity. Since its inception, over 30,000 companies have registered.

11. Marjorie Coeyman, "ISO 9000 Gaining Ground in Asia/Pacific," *Chemical Week*, April 28, 1993, p. 54. In some parts of the Asia/Pacific region, the ISO 9000 quality standards have almost become domestic standards.

12. In a 1993 conference of the National Society of Professional Engineers, the topic of compliance with the EC's ISO 9000 quality control standards was discussed. See Jane C. Edmunds, "Engineers Want Quality," *ENR*, Feb. 8, 1993, p. 15. For Power Transmission Distributors (The Association), see Beate Halligan, "ISO Standards Prepare You to Compete," *Industrial Distribution*, May 1992, p. 100. The concern of the public utilities industry is reported in Greg Hutchins, "ISO Offers a Global Mark of Excellence," *Public Utilities Fortnightly*, April 15, 1993, p. 35. The computer industry's concern is reflected in Gary H. Anthes, "ISO Standard Attracts U.S. Interest," *Computerworld*, April 26, 1993, p. 109.

13. "Support Group Formed for Companies Seeking ISO 9000," *Industrial Engineering*, March 1993, p. 8. The National ISO 9000 Support Group will provide information, support, advice, and training at low cost to any American company interested in the ISO 9000 process. The goal of the group is to allow the free exchange of information and questions between companies seeking ISO registration.

14. Copies of the standards are available for a small fee from the American Society of Quality Control (Tel. 414-272-8575) and the American National Standards Institute (Tel. 212-642-4900).

15. Donald W. Marquardt, "ISO 9000: A Universal Standard of Quality," *Management Review*, Jan. 1992, p. 51. See also Kymberly K. Hockman, "The Last Barrier to the European Market," *Wall Street Journal*, Oct. 7, 1991, Section A, p. 14.

16. Michael E. Raynor, "ISO Certification," *Quality*, May 1993, pp. 44–45.

17. Adapted from John D. Flister and Joseph J. Jozaitis, "PPG's Journey to ISO 9000," *Management Accounting*, July 1992, p. 34.

18. Adapted from Kymberly K. Hockman and David A. Erdman, "Gearing Up for ISO 9000 Registration," *Chemical Engineering*, April 1993, p. 129.

19. Adapted from *Business Week*, Oct. 19, 1992 and ASQC *ANSI/ASQC Q-90-1987— Quality Management and Quality Assurance Standards*.

20. Gary H. Anthes, "ISO Standard Attracts U.S. Interest," *Computerworld*, April 26, 1993, p. 109.

21. Donald W. Marquardt, "ISO 9000: A Universal Standard of Quality," *Management Review,* Jan. 1992, p. 52.

22. See Elisabeth Kirschner, "Nalco: Registration in Context," *Chemical Week,* April 28, 1993, p. 71. See also Marjorie Coeyman, "FMC: The Benefits of Documentation," *Chemical Week,* April 28, 1993, p. 69.

23. See Giovanni Grossi, "Quality Certifications," *Internal Auditor,* Oct. 1992, p. 33–35 and Gary M. Stern, "Sailing to Europe: Can Auditing Play a Role in the New International Quality Standards?" *Internal Auditor,* Oct. 1992, pp. 29–33. Both of these authors, who are members of the internal auditing profession, argue for an expanded role for internal auditors in the certification and follow-on process.

24. "IBM, Help/Systems Receive ISO Certification," *Systems 3X-400,* Feb. 1993, p. 16. ABS won the Baldrige Quality Award in 1991.

25. Emily S. Plisher, "Seeking Recognition: U.S. Auditors Build Their Base," *Chemical Week,* Nov. 11, 1992, pp. 30–33. See also a special report entitled "Confusion Persists on Issue of Registrar Accreditation," *Chemical Week,* April 28, 1993, p. 42. As of April 1993, ANSI/RAB had accredited 27 quality system registrars. The "unofficial" list is contained in this endnote citation.

26. Gary Spizizen, "The ISO 9000 Standards: Creating a Level Playing Field for International Quality," *National Productivity Review,* Summer 1992, p. 335.

27. This is the estimate of OTS Registrars of Houston, an ISO 9000 registrar. "Small Companies Are Finding it Pays to Think Global; Firms Win New Business by Adopting International Quality Standards," *Wall Street Journal,* Nov. 19, 1992, Section B, p. 2.

28. Donald W. Marquardt, "ISO 9000: A Universal Standard of Quality," *Management Review,* Jan. 1992, p. 51. See also Ian Hendry, "ISO Standardizes Quality Efforts," *Pulp & Paper,* Jan. 1993, p. S4. Several of these firms report a cost of certification of about $112,000. Also, General Chemical's Green River plant achieved certification at an estimated cost of $150,000. Rick Mullin, "General Registers Green River Site: First ISO 9002 for Natural Soda Ash," *Chemical Week,* April 28, 1993, p. 59.

17

WHAT IS
the BALDRIGE AWARD?

Quality has come a long way since renewed interest began in the middle and late 1970s. Prior to that time, many people considered the emphasis on quality as just one more passing phase in a string of business fads—value analysis...management by objective...Theory X and Y...portfolio management...and so on. The impact of the Baldrige Award has laid to rest any notion that quality is not here to stay.

There is little doubt that quality will remain the major competitive issue of the 1990s and beyond. Increasing global competition and customer sensitivity have given quality increasing visibility. An additional impetus was provided when Congress established the Baldrige Award in 1987 as a result of Public Law 100-107. Background information on the law mentions foreign competition as the major rationale. No other business prize or development in management theory can match its impact. As evidence of this impact, over twenty states are working to develop regional quality programs.[1]

The award has set a national standard for quality, and hundreds of major corporations use the criteria in the application form as a basic management guide for quality improvement programs. Although the award has its detractors,[2] it has effectively created a new set of standards—a benchmark for quality in U.S. industry.

Applicants must address seven specific categories. These categories of examination items and their respective point values are listed in Table 17-1. The Baldrige Award framework and the dynamic relationships among the criteria are shown in Figure 17-1.

Table 17-1 1993 Examination Items and Point Values[3]

Examination categories/items	Point values
1.0 Leadership	**95**
Senior executives' *personal* leadership and involvement in creating and sustaining a customer focus and clear and visible quality values. Also examined is how the quality values are integrated into the company's management system and reflected in the manner in which the company addresses its public responsibilities and corporate citizenship.	
1.1 Senior executive leadership	45
1.2 Management for quality	25
1.3 Public responsibility and corporate citizenship	25
2.0 Information and Analysis	**75**
The scope, validity, analysis, management, and use of data and information to drive quality excellence and to improve operational and competitive performance. Adequacy of company data, information, and analysis system to support improvement of the company's customer focus, products, services and internal operations.	
2.1 Scope and management of quality and performance data and information	15
2.2 Competitive comparisons and benchmarking	20
2.3 Analysis and uses of company-level data	40
3.0 Strategic Quality Planning	**60**
The planning process and how all key quality requirements are integrated into overall business planning. The company's short- and longer-term plans and how quality and operational performance are deployed to all work units.	
3.1 Strategic quality and company performance planning process	35
3.2 Quality and performance plans	25
4.0 Human Resource Development and Management	**150**
The key elements of how the work force is enabled to develop its full potential to pursue the company's quality and operational performance objectives. Also examined are the company's efforts to build and maintain an environment for quality excellence conducive to full participation and personal and organizational growth.	
4.1 Human resource planning and management	20
4.2 Employee involvement	40
4.3 Employee education and training	40
4.4 Employee performance and recognition	25
4.5 Employee well-being and satisfaction	25

Table 17-1 (continued) 1993 Examination Items and Point Values[3]

Examination categories/items	Point values
5.0 Management of Process Quality	140
Systematic processes the company uses to pursue ever-higher quality and company operational performance. The key elements of process management, including R&D, design, management of process quality for all work units and suppliers, systematic quality improvement, and quality assessment.	
5.1 Design and introduction of quality products and services	40
5.2 Process management: product and service production and delivery processes	35
5.3 Process management: business processes and support services	30
5.4 Supplier quality	20
5.5 Quality assessment	15
6.0 Quality and Operational Results	180
The company's quality levels and improvement trends in quality, company operational performance, and supplier quality. Current quality and operational performance levels relative to those of competitors.	
6.1 Product and service quality results	70
6.2 Company operational results	50
6.3 Business process and support service results	25
6.4 Supplier quality results	35
7.0 Customer Focus and Satisfaction	300
The company's relationships with customers and its knowledge of customer requirements and of the key quality factors that drive marketplace competitiveness. Also the company's methods to determine customer satisfaction, current trends and levels of customer satisfaction and retention, and these results relative to competitors.	
7.1 Customer expectations: current and future	35
7.2 Customer relationships management	65
7.3 Commitment to customers	15
7.4 Customer satisfaction determination	30
7.5 Customer satisfaction results	85
7.6 Customer satisfaction comparison	70
Total Points	**1000**

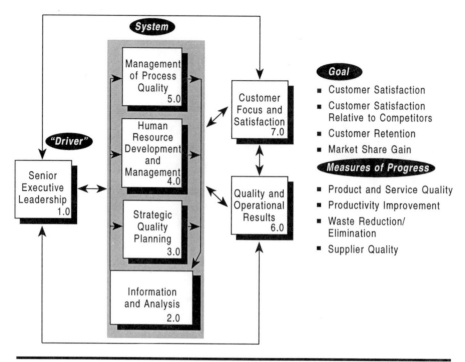

Figure 17-1 Baldrige Award Criteria Framework: Dynamic Relationships

Meeting the criteria is not an easy matter. A perfect score is 1000. The distribution of scores for the 203 applicants during the first three years (1988, 1989, 1990) is shown in Table 17-2. Of the 1203 applicants, only 9 were selected for the award.

An indication of the interest in the Baldrige is the number of application guidelines (167,000 in 1990) requested. In the first 3 years, 203 companies applied and 9 won: 6 manufacturers, 2 small companies, and 1 service company (Federal Express). Winners of the award are required to share their successful strategies with other companies. IBM's Rochester, Minnesota site, home of the Applications System/400 and a 1990 winner, attributes the success of the division to the way in which it appropriated the ideas of Motorola, Xerox, and Milliken, winners in prior years. This sharing of ideas is a central purpose of the National Institute of Standards and Technology, the administering agency.[14] The sharing policy by winners ensures a multiplier effect.

Another indication of the award's leverage is the stringent criteria

Table 17-2 Distribution of Scores

Scoring range	Number of applications		
	1988	1989	1990
0–125	0	0	0
126–250	0	1	7
251–400	1	8	18
401–600	31	15	51
601–750	23	12	19
751–875	11	4	2
876–1000	0	0	0
Total	66	40	97

related to quality assurance for products and services purchased by external providers (suppliers) of goods and services. It is clear that suppliers are a critical link in the chain of processes that constitute TQM. As a result, many companies require their suppliers to apply for the Baldrige. For example, Motorola and Westinghouse, two winners, will not do business with a supplier that has not applied for the award and does not use its criteria. Another winner, Globe Metallurgical, is certified as a supplier by Ford. Globe in turn requires certification by its suppliers. Thus, the number of firms using the Baldrige criteria may grow geometrically as first-tier suppliers certify second-tier suppliers and so on.

Hewlett-Packard, IBM, Motorola, Westinghouse, and 3M are among the many companies that use the application as a guide for managers and a checklist for internal quality standards:

■ But basically, the Baldrige criteria will be the way we judge our own operations from now on. The reason is simple: The Baldrige Award process is a basic blueprint on how to do the quality process.[5]

> 3M

■ Competing for the award motivated people to a level I didn't think possible.[6]

> General Manager
> GM Cadillac Division

■ The National Quality Award process enabled the company to look at itself through the eyes of the customer, and every aspect of the business came under scrutiny.[7]

Xerox

■ Managers of IBM's Santa Teresa, California lab are required to score their operations every 90 days using the criteria.

The winners for the five-year period since the beginning of the award in 1988 are shown in Table 17-3.[8]

Table 17-3 Award Winners: 1988 to 1992

1992 Award Winners	1990 Award Winners
Manufacturing AT&T Network Systems Group Transmission Systems Business Unit Morristown, NJ	**Manufacturing** Cadillac Motor Car Company Detroit, MI
Texas Instruments, Inc. Defense Systems & Electronics Group Dallas, TX	IBM Rochester Rochester, MN
Service AT&T Universal Card Services Jacksonville, FL	**Service** Federal Express Corp. Memphis, TN
The Ritz-Carlton Hotel Company Atlanta, GA	**Small Business** Wallace Co., Inc. Houston, TX
Small Business Granite Rock Company Watsonville, CA	**1989 Award Winners**
1991 Award Winners	**Manufacturing** Milliken & Company Spartanburg, SC
Manufacturing Solectron Corp. San Jose, CA	Xerox Business Products and Systems Stamford, CT
Zytec Corp. Eden Prairie, MN	**1988 Award Winners**
Small Business Marlow Industries Dallas, TX	**Manufacturing** Motorola, Inc. Schaumburg, IL
	Westinghouse Commercial Nuclear Fuel Division Pittsburgh, PA
	Small Business Globe Metallurgical, Inc. Cleveland, OH

ENDNOTES

1. Curt W. Reimann, "America Unites Behind the Baldrige Quality Crusade," *Electronic Business,* October 15, 1990, p. 63. Reimann is Director of the Malcolm Baldrige National Quality Award and Associate Director for Quality Programs at the National Institute of Standards and Technology, the agency that administers the Baldrige Award program. A good summary of what it takes to compete for the Baldrige is contained in Curt Reimann, "Winning Strategies for Quality Improvement," *Business America,* March 25, 1991, pp. 8–11. See also "A Standard for All Seasons," *Executive Excellence,* March 1991, p. 9, and "The Baldrige Award: Leading the Way to Quality," *Quality Progress,* July 1989, pp. 35–39.

2. See Jeremy Main, "Is the Baldrige Overblown?" *Fortune,* July 1, 1991, pp. 62–65. Philip Crosby of *Quality Is Free* fame scorns the paperwork, thinks that customers rather than the company applying should do the nominating, and deplores the lack of financial measures. Tom Peters, co-author of *Search for Excellence,* complains that the criteria are "strangely silent on the subject of bureaucracy." There was also a bit of sour grapes when Cadillac won the award in 1990.

3. A more detailed description is contained in The 1993 Award Criteria and The 1993 Application Forms and Instructions. These two documents can be obtained from Malcolm Baldrige National Quality Award, National Institute of Standards and Technology, Route 270 and Quince Orchard Road, Administration Building, Room A537, Gaithersburg, MD 20899 (Tel. 301-975 2036).

4. Michael Fitzgerald, "Quality: Take it to the Limit," *Computerworld,* Feb. 11, 1991, pp. 71–78.

5. Remarks of A. G. Jacobson in a presentation to the Conference Board Quality Conference, April 2, 1990.

6. Jeremy Main, "Is the Baldrige Overblown?" *Fortune,* July 1, 1991, p. 63.

7. Company brochure entitled "The Xerox Quest for Quality and the National Quality Award."

8. Texas Instruments is one of the several Baldrige winners that attribute their turnaround to the adoption of the principles of TQM. See *Fortune,* Nov. 30, 1992, p. 80–83. In October 1992 the Ritz-Carlton Hotel Company became the first hotel pcompany to win the Baldrige Award. Their approach to quality relies on traditional TQM principles. Edward Watkins, "How Ritz-Carlton Won the Baldrige Award," *Lodging Hospitality,* Nov. 1992, pp. 22–24. In 1990, AT&T chairman and CEO Robert Allen created the Chairman's Quality Award, the criteria and examination process for which were taken from the Malcolm Baldrige Award. See Rick Whiting, "AT&T Started a Quality Bonfire to Learn How to Put it Out," *Electronic Business,* Oct. 1992, pp. 95–103.

18

REENGINEERING

Customer satisfaction has increasingly become the cardinal principle governing any successful business. In some cases, marketing campaigns have been reformulated and new slogans invented to take advantage of the impact of advertising on customers. Despite the new emphasis on customer satisfaction by companies, there has been a high incidence of complaints, anger, rage, and acute disappointment over products. One of the reasons for this lies in the fact that the old ways and processes have become so severely inadequate that the mere realignment of old values is no longer acceptable. In many cases, the present system can no longer be fixed and incremental improvements are not sufficient. What is needed is the reengineering of the entire system.

WHAT IS REENGINEERING?

According to Hammer,[1] reengineering may be defined as "the fundamental rethinking and radical design processes to achieve dramatic improvements in critical contemporary measures of performance such as cost, quality, service, and speed." Reengineering does not strive to revamp an existing process. It seeks to enhance the celerity of the delivery of a product without compromising its quality by improving the utilization of materials, labor, and equipment. Janson[2] notes that "By focussing on making improvements in all dimensions of the service organization—human dimension, work process dimension, and the technological dimension—reengineering helps companies overcome systematic work barriers that interfere with efforts to achieve higher levels of customer satisfaction." For some other proponents of reengineering, it seems to be the utilization of computers and modern technology to enhance the work process. Lawrence[3] states that, "It involves redesigning business processes to take advantage of the enormous potential of computer and information technology." In order for companies to embrace the concept of reengineering, they must be able to break away from previously followed conventional rules and policies and be open to changes that would make their businesses more productive. Hammer[4] states that "Reengineering strives to break away from the old rules about how we organize and conduct business. It involves recognizing and rejecting some of them and finding imaginative new ways to accomplish work."

REENGINEERING as APPLIED to ANY BUSINESS PROCESS

Hammer[1] states that business process may be defined as a set of logically related tasks performed to achieve a defined business outcome. A set of processes forms a business system—the way in which a business unit, or a collection of units, carries out its business.

Assuming a company has decided that its process are ineffective and inefficient, the following are the five major steps the company should embark on to redesign its process, according to Hammer:[4]

- **Develop business vision and process objectives.** This step involves prioritizing objectives and setting targets for the future.
- **Identify processes to be redesigned.** This involves identifying critical or bottleneck processes and envisioning steps to avert shortcomings in them.
- **Understand and measure existing processes.** This involves identifying current problems and setting a baseline.
- **Identify information technology levels.** This involves bringing those involved in the process to a brainstorming session to identify new approaches.
- **Design and build a prototype of the process.** This includes implementing organizational and technical aspects.

The ESSENCE of REENGINEERING

At the heart of reengineering, according to Hammer,[4] is the notion of discontinuous thinking of organizing and breaking away from the outdated rules and fundamental assumptions that underlie operations. Unless these rules are changed, breakthroughs in performance are extremely difficult. Old assumptions must be challenged, and the old rules that made the business underperform in the first place must be shed. Most contemporary businesses run on the basis of policies laid down during past decades. These assumptions about technologies, people, and organizational goals probably are no longer valid. Quality and customer service are increasingly becoming the primary focus of any company. A large portion of the population is educated and capable of assuming responsibility, and workers cherish their autonomy, expect to have a say in how the business is run, and demand quality. Consequently, the present business process and structures are outmoded and obsolete. Work structures and processes have not kept pace with changes in technology, demographics, and business objectives. Systems for imposing control and discipline on those involved in work stem from the post-war period. Conventional process structures are fragmented, piecemeal, and myopic. Consequently, employees substitute the narrow goals of their particular department for the larger goals of the process as a whole. Moreover, when work is turned over from one person to another, delays and errors become increasingly inevitable. Managers have tried to break loose from outmoded business processes and the design principles that underlie them. Reengineering seeks to provide a new perspective to business operations and processes. According to Hammer and Champy,[1] *"Reengineering* requires looking at the fundamental processes of the business from a cross-functional perspective."

Ford discovered that reengineering the accounts payable department alone was futile. The appropriate focus of the effort was what might be called the goods acquisition process, which includes purchasing and receiving as well as accounts payable. According to Hammer,[4] one way to ensure that reengineering has a cross-functional perspective is to assemble a team that represents the functional units involved in the process being reengineered and all the units that depend on it. Rather than looking for opportunities to improve the current process, the team should determine which of its processes really add value and search for new ways to achieve the end result.

In short, reengineering efforts strive for dramatic levels of improvement. They break away from conventional wisdom and the constraints of organizational boundaries and are broad and cross-functional in scope. They use information technology not to automate an existing process but to create a new one.

PRINCIPLES of REENGINEERING

Creating new rules tailored to the modern environment ultimately requires a new conceptualization of the business process. However, according to Hammer,[4] reengineering need not be haphazard. Some of the principles that companies have already discovered while reengineering their business processes can help jump start the effort for others.

Organize around outcomes, not tasks. This principle calls for the use of one person to perform all the steps in a process. Design that person's job around an objective or outcome instead of a single task. The following is an example of an electronics company which had separate organizations performing each of the five steps between selling and installing equipment.

> ▪ One group determined customer requirements, another translated those requirements into internal product codes, a third conveyed that information to various plants and warehouses, a fourth received and assembled the components, and a fifth delivered and installed the equipment. The customer order moved systematically from step to step, but this sequential processing caused problems. The people getting the information from the customer in step 1 had to get all the data needed throughout the process, even if it was not needed until step 5. In addition, the many handoffs were responsible for many errors. Finally, any complaints from customers were referred all the way back to step 1, which caused inordinate delays in customer service. When the company ultimately reengineered, the assembly line approach was eliminated. Responsibility for the various steps was compressed and assigned to one person, the customer service representative. This person would oversee the entire process. The customer service representative expedites and coordinates the process, much like a general contractor.

Have those who use the output of the process perform the process. In an effort to capitalize on the benefits of specialization and scale, many organizations established specialized departments to handle specialized processes. Computer-based data and expertise are now more readily available, enabling departments, units, and individuals to do more for themselves. Opportunities exist to reengineer processes so that individuals who need the result of the process can perform it themselves. When people closest to the process perform it, there is little need for the overhead associated with managing it. Interfaces and liaisons can be eliminated, as can the mechanisms used to coordinate those who perform the process with those who use it. Moreover, the problem of capacity planning for those who perform the process is greatly reduced.

Subsume information processing work into the real work that produces the information. The previous two principles were used to compress linear processes. This principle suggests moving work from one person or department to another. Most companies establish units which do nothing but collect and process information created by other departments. This arrangement reflects the old rule about specialized labor and the belief that people at lower levels are incapable of acting on information they generate. An accounts payable department collects information from purchasing and receiving and reconciles it with data provided by the vendor. Quality assurance gathers and analyzes information received from production. Redesigning the accounts payable process embodies the new rule, wherein receiving (which produces the information about the goods received) processes this information instead of sending it to accounts payable. The new system can easily compare the delivery with an order and initiate appropriate action.

Treat geographically dispersed resources as though they were centralized. The conflict between centralization and decentralization is that decentralizing a resource gives better service to those who use it, but at the cost of abundance and missed economies of scale. Companies no longer have to make such trade-offs. They can use databases, telecommunication networks, and standard processing systems to realize the benefits of scale and coordination while maintaining the benefits of flexibility of service.

Link parallel activities instead of integrating their results. This principle seeks to forge the links between functions and to coordinate them while their activities are in process rather than after they have been completed. Communication networks, shared databases, and teleconferencing can bring independent groups together so that coordination is ongoing.

Put the decision point where the work is performed, and build control into the process. In most organizations, those who do the work are distinguished from those who monitor the work and make decisions about it. The tacit assumption is that the people actually doing the work have neither the time nor the inclination to monitor and control the work and therefore lack the knowledge and scope to make decisions about it. The entire hierarchical management structure is built on this assumption. Accountants, auditors, and supervisors check, record, and monitor work. Managers handle any exceptions. The new principle suggests that the people who do the work should make decisions and that the process itself can have built-in controls. Pyramidal management layers can therefore be compressed and the organization flattened.

Capture information once and at the source. This last rule is simple. When information was difficult to transmit, it made sense to collect it repeatedly. Each person, department, or unit had its own requirements and forms. Companies simply had to live with the associated delays, entry errors, and costly overhead. However, by integrating and connecting these systems, the company was able to eliminate this redundant data entry, along with the attendant checking functions and the seemingly inevitable errors.

The THREE R'S of REENGINEERING

The length and difficulty of the engineering process vary greatly among organizations, depending on the need for change and the extent of employee involvement. Janson[2] states that every reengineering effort involves three basic phases:

1. **Rethink:** This phase of the reengineering improvement process requires examining the organization's current objectives and underlying assumptions to determine how well they incorporate the renewed commitment to customer satisfaction. Another valuable exercise in this phase is to examine the critical success factors— those areas in which the organization clearly stands apart from the competition. Do they contribute to the new customer satisfaction goals?

2. **Redesign:** This phase of the reengineering improvement process requires an analysis of the way the organization produces the products or services it sells—how jobs are structured, who accomplishes what tasks, and the results of each procedure. Then, a determination must be made as to which elements should be redesigned to make jobs more satisfying and more customer focused.

3. **Retool:** This phase of the reengineering process requires a thorough evaluation of the current use of advanced technologies, especially electronic word and data processing systems, to identify opportunities for change that can improve quality of service and customer satisfaction.

REENGINEERING in the SERVICE INDUSTRY

Spadaford[5] notes that in the service sector, despite speed and courtesy in addressing customers, customer hostility persists due to the inability of the service provider to maintain consistency in delivery. Another frequent problem is delay, due to the inability of the employee to make pertinent and satisfying decisions when confronted by an impatient customer. Sufficient evidence of this is found in banks and fast-food restaurants. However, some companies which start anew seem to better satisfy their customers in their incipience than a few years later. One reason is that their operations are simpler and smaller. As they grow, companies become increasingly intense and complex, and work habits evolve into unacceptable levels of performance. To remain competitive today, service organizations need to focus on customer satisfaction and on real customer needs and expectations. They need to operate according to the standards of the customers.

According to Spadaford,[5] the contributing factors in the decline of quality in the service industry are as follows:

- The business grew in complexity and scale. The current process was not truly designed, but rather was created piecemeal in response to problems.
- Organizational fragmentation created barriers to cooperation.
- A 21-year-old loan system could not fully accommodate loan servicing needs.

▪ Work flows required multiple handoffs on different floors in an extended assembly line, with much finger pointing for delays or errors.

▪ Morale was never high because no one involved in the service process ever felt like a winner.

▪ With so many hands in the process, no one was fully accountable for delivering service to the customer.

According to Janson,[2] the concept of reengineering holds a significant promise for the service sector. The following are some of the salient features:

1. **Make the customer the starting point for change.** Companies that reengineer start to rebuild themselves from the outside in, without taking into consideration how they "normally" operate. This means identifying what the customers really want and then creating the kinds of jobs and organizational structures that can satisfy those expectations. Another way of doing this is to realign jobs by type of customer. Some companies create work teams that face off against a specific geographic region or against a particular market segment. This builds a strong relationship with customers and helps workers better understand customer needs.

2. **Design work processes in light of organizational goals.** Companies that design work processes according to organizational goals become more focused toward the customer. Organizations that reengineer often make drastic changes in existing jobs by integrating work procedures or tasks and empowering workers with more authority and responsibility. Application of this principle results in a positive attitude toward the customer and a subsequent breakdown of departmental distinctions.

3. **Restructure to support front-line performance.** In a customer-focused environment, every aspect of the organization strives to promote the highest level of service to its customers, especially those who come in direct contact with customers. Consequently, organizations that undertake reengineering build work teams to support their customer service representatives or create "work station professionals," who can perform both front and back office functions. Application of this principle generally results in better designed jobs, with increased accountability, greater skill and task variety, and a clear customer focus.

The IMPACT of REENGINEERING on the SERVICE INDUSTRY

According to Janson,[2] reengineering is not a single technique. It represents a major advance over conventional management strategies for improvement. As an integrated approach, it involves three dimensions of a service organization:

1. **The human dimension:** To achieve a stronger customer focus, workers at all levels must readjust their thinking and recognize that customer satisfaction is the overriding goal. Some companies that reengineer achieve this by rewriting their mission statement to reflect the primacy of the customer or by promoting a new vision to reinforce the central role that customer satisfaction now plays. Other companies engage in formal training to help employees become better listeners, probe for customer concerns more effectively, or satisfy customer needs more creatively. The motive in reengineering is to become more motivated to provide superior service and be skilled at doing it.

1. **The work process dimension:** Work systems must be designed not according to their internal logic or any external definition of efficiency, but according to how well they satisfy customer needs. This sometimes requires substantial structural changes in an organization—changes that do more than just revamp job descriptions. It may mean setting up work teams to perform all the functions once divided among several departments or combining several individual jobs to create one multiskilled customer service professional. In every case, total reevaluation of management's role in the organization comes into play, and lower level workers typically assume far greater responsibility for service quality.

3. **The technology dimension:** New technologies should be introduced not only because they are more advanced, but because they truly support the organization in its drive to achieve higher levels of customer satisfaction. Distribution access computer systems, for example, can lessen the need for callbacks and transfers and give those who actually service the customer easier access to important information. Most importantly, technology should be used to automate secondary work functions, leaving service workers free to concentrate on more critical matters, such as satisfying customer needs and solving problems.

QUALITY and REENGINEERING

According to Spadaford,[5] the tools of quality always come in handy when laying out procedures for reengineering. Quality can be used to define a series of internal measures of process outcomes that are important to customers. Standards and goals should be established for each of these measures, and tracking methods for standards as an ongoing process can be extremely beneficial. Weekly informal group meetings along with a monthly meeting could be an ideal tool to review measures for compliance. In addition, periodic analysis of functions from the standpoint of value would result in identification of areas that need to be eliminated. This also aids the department in the redeployment of resources when faced with frequently changing market requirements. Reengineering as a tool shifts its focus from a leadership model of command and control to one of commitment and trust. It is a challenge to attain this environment in a bank because of the high-risk transactions and rigorous controls. To benefit the newly developed organizations and for the continuing success of their endeavors, managers must demonstrate the following:

1. Set a clear strategic direction and demonstrate commitment to it.

2. Openly communicate with staff, especially when dealing with bad news or problems.

3. Encourage staff involvement whenever possible, so that managers have a say in how their jobs are affected by any changes.

4. Experiment with new ideas and concepts and reward not only results, but also efforts to bring about a positive change.

5. Weigh the risks of change against the returns expected and expend effort only on projects that provide clear long-term payoffs.

6. Withhold the expectation of a quick realization of benefits, as many changes will take years to truly blossom.

7. Focus more on outcomes and less on activities. This will create a mindset for adding value and reducing marginally useful activities.

8. Place primary emphasis on creating the appropriate climate for people to attain excellent performance and personal growth.

Through high personal standards and the staff empowerment that results from these leadership attributes, a high-performance work unit will flourish.

BENEFITS of REENGINEERING

Rohm[6] proposes some important benefits of reengineering:

- **Revolutionary thinking:** Reengineering encourages organizations to abandon conventional approaches to problem solving and to "think big."
- **Breakthrough improvement:** The slow, cautious process of incremental improvement leaves many organizations unprepared to compete in today's rapidly changing marketplace. Reengineering helps organizations make noticeable changes in the pace and quality of their response to customer needs.
- **Organizational structure:** Through reengineering, an insurance company, for example, was transformed from a factory-style organization that was essentially rule driven and job centered to a marketing organization that focuses directly on the customer. The current primary objective of the organization is to identify real customer needs, rather than create products that ignore the needs and wants of the customers.
- **Organizational renewal:** Reengineering often results in radically new organizational designs that can help companies respond better to competitive pressures, increase market share and profitability, and improve cycle times, cost ratios, and quality.
- **Corporate culture:** Perhaps the major accomplishment of the reengineering effort is the change that occurs in the corporate culture and in the basic principles by which departments operate. Workers at all levels are encouraged to make suggestions for improvement and to believe that management will listen to what they have to say. However, sometimes management must still struggle to draw people into the participative process. Reengineering will eventually help the culture in the organization to evolve from an insular one to one that accepts change and knows how to deal with it.
- **Job redesign:** Reengineering has helped create more challenging and more rewarding jobs with broader responsibilities for employ-

ees. Workers who are used to performing only one simple task over and over are now involved in the entire process of prospecting for customers, making a sale, and processing and submitting applications.

ENDNOTES

1. M. Hammer and J. Champy, "The Promise of Reengineering," *Fortune,* May 3, 1993, pp. 94–97.
2. R. Janson, "How Reengineering Transforms Organizations to Satisfy Customers," *National Productivity Review,* Winter 1992/94, pp. 45–53.
3. P. J. Lawrence, "Reengineering the Insurance Industry," *Best's Review,* May 1991, pp. 68–75.
4. M. Hammer, "Reengineering Work: Don't Automate, Obliterate," *Harvard Business Review,* July–Aug. 1990, pp. 104–112.
5. J. F. Spadaford, "Reengineering Commercial Loan Servicing at First Chicago," *National Productivity Review,* Winter 1992/93, pp. 65–72.
6. C. E. Rohm, "The Principal Insures a Better Future by Reengineering Its Individual Insurance Department," *National Productivity Review,* Winter 1992/93, pp. 55–64.

BIBLIOGRAPHY

1. J. E. Layden, "Reengineering the Human/Machine Partnership," *Industrial Engineering,* April 1993, p. 14.
2. T. J. McCabe and E. S. Williamson, "Tips on Reengineering Redundant Software," *Datamation,* April 15, 1992, pp. 71–74.
3. G. Rifkin, "Reengineering Aetna," *Forbes ASAP,* pp. 78–83.
4. K. Hayley, J. Plewa, and M. Watts, "Reengineering Tops CIO Menu," *Datamation,* April 15, 1993, pp. 73–74.
5. J. Kador, "Reengineering to Boost Software Productivity," *Datamation,* Dec. 15, 1992, pp. 57–58.
6. S. L. Huff, "Reengineering the Business," *National Productivity Review,* Winter 1992, pp. 38–42.
7. T. C. Taylor, "Plugging into the Future," *Sales & Marketing Management,* June 1993, pp. 20–24.

IV

CASES in QUALITY

In Part IV, three comprehensive cases are presented: one from the manufacturing sector (Zytec—a Baldrige Award winner) and two from the service sector (a health care facility and Liberty Bank). The Liberty Bank case is used by award administrators to train on-site examiners in companies applying for the Baldrige Award. All three cases serve as vehicles to illustrate good and poor TQM practices.

19

ZYTEC

OVERVIEW

Zytec designs and manufactures electronic power supplies and repairs power supplies and CRT monitors. Power supply customers are original equipment manufacturers who integrate Zytec power supplies into their systems, which range from computers to peripherals, medical equipment, office products, and test equipment. Repair customers are original equipment manufacturers, third-party maintenance companies, and large end users.

Zytec sells, by a direct sales force, to customers in the United States. However, since most customers are large multinational companies, many of Zytec's power supplies are exported around the world. Zytec competes for business with Far East and European companies as well as approximately 400 U.S. companies.

Zytec is currently the fifth largest multiple output switching power supply company in the U.S., the fastest growing U.S. electronic power supply company, and the largest power supply repair company in North America.

Zytec is a small, employee-owned company, with approximately 800 employees located at its Minnesota headquarters in Eden Prairie and manufacturing facility in Redwood Falls.

Since the company began in 1984, Zytec has used the quality and reliability of its products and services as the key strategy to differentiate it from the competition. The following abstract of Zytec's 1991 applications for the Minnesota Quality Award and the Malcolm Baldrige National Quality Award outlines the scope and success of that strategy.

Reprinted with permission of Zytec U.S.A., Redwood Falls, MN, a 1991 Malcolm Baldrige National Quality Award winner.

CATEGORY ONE: LEADERSHIP

When Zytec started in January 1984, it chose three words as the new company's focus: *Quality, Service,* and *Value.*

> "Zytec is a company that competes on value; it provides technical excellence in its products and believes in the importance of execution.
>
> We believe in a simple form and a lean staff, the importance of people as individuals, and the development of productive employees through training and capital investment.
>
> We focus on what we know best, thereby making a fair profit on current operations to meet our obligations and perpetuate our continued growth."

To carry out this mission, Zytec's senior executives decided to embrace Dr. W. Edwards Deming's 14 Points for Management as the cornerstone of the company's quality improvement culture. They established the Deming Steering Committee to guide the Deming process and championed individual Deming Points, acted as advisors to the three Deming Implementation Teams, and developed the Zytec Total Quality Commitment statement.

Meetings were held with every Zytec employee to increase knowledge of Deming's Points, and many employees have attended half-day, two-day, and four-day Deming seminars. As a result, Deming's 14 Points for Management guide Zytec's actions, from long-range strategic planning to employee empowerment to leadership. Their effect is visible throughout this document.

In 1988, Zytec's CEO and the vice president of marketing and sales visited Japan on a TQC study mission. As a result of this trip, Zytec initiated a process called Management By Planning (MBP). The process involves employees in establishing long-range plans and short-term objectives.

Dr. Deming's 14 Points:
Methods for Management of Productivity and Quality

1. Create constancy of purpose toward improvement of product and service, with the aim to become competitive and to stay in business, and to provide jobs.

2. Adopt a new philosophy. We are in a new economic age. Western management must awaken to the challenge, must learn their responsibilities, and take on leadership for change.

3. Cease dependency on inspection to achieve quality. Eliminate the need for inspection on a mass basis by building quality into the product in the first place.

4. End the practice of awarding business on the basis of price tag. Instead, minimize total cost. Move toward a single supplier for any one item, on a long-term relationship of loyalty and trust.

5. Improve constantly and forever the system of production and service, to improve quality and productivity and thus constantly decrease costs.

6. Institute training on the job.

7. Institute leadership (see point 12). The aim of leadership should be to help people and machines and gadgets to do a better job. Leadership of management is in need of overhaul, as well as leadership of production workers.

8. Drive out fear, so that everyone may work effectively for the company.

9. Break down barriers between departments. People in research, design, sales, and production must work as a team to foresee problems of production as well as after-sale use.

10. Eliminate slogans, exhortations, and targets for the work force, asking for zero defects and new levels of productivity.

11. (a) Eliminate work standards (quotas) on the factory floor. Substitute leadership. (b) Eliminate management by objective. Eliminate management by numbers and numerical goals. Substitute leadership.

12. (a) Remove barriers that rob hourly workers of their right to pride of workmanship. The responsibility of supervisors must be changed from sheer number to quality. (b) Remove barriers that rob people in management and in engineering of their right to pride of workmanship. This means, inter alia, abolishment of the annual or merit rating and of management by objective, management by numbers.

13. Institute a vigorous program of education and self-improvement.

14. Put everybody in the company to work to accomplish the transformation. The transformation is everybody's job.

The MBP process, described in Category Three of the Baldrige Criteria, has become a major driver of quality values within Zytec. Zytec's senior executives set corporate objectives to guide the process, then lead the six cross-functional teams that review and develop individual plans, by key functional areas, for presentation to employees.

For 1991, Zytec's corporate objectives are to:

▪ Improve the quality of our products and processes to become a Six Sigma company by 1995.

▪ Reduce our total cycle time.

▪ Improve our service to our customer.

▪ Improve Zytec so we win the Malcolm Baldrige National Quality Award by 1995.

Each department manager, working with his/her people, develops a supporting objective for each of the corporate objectives. Nearly every employee is on an MBP team working on at least one supporting objective. The CEO plays "catch ball" with each department to coordinate establishing their objectives, monthly goals, and action plans. During monthly operations review meetings, senior management reviews progress on the MBP corporate objectives. Each of Zytec's 33 departments presents its progress on its four objectives. If a department needs help, a senior executive is assigned responsibility for providing whatever is needed to improve the process.

MBP is one way senior management communicates Zytec's quality mission, goals, and objectives throughout the company. Senior executives also participate in quarterly communication meetings with all employees, during which they discuss what is happening in the company and marketplace and answer questions. All employees receive full financial information from their managers five days after the end of each month.

Senior management is actively involved in quality training. In 1988, Zytec's senior management developed and taught a 16-hour training class on culture, difficulty of change, customer service, quality, communication, and leadership to all employees. In 1990, the vice president of marketing and sales led his department in developing a "Service America" training class, which he taught to all Zytec employees.

Senior executives meet often with customers and suppliers and participate in benchmarking competitors and world-class companies. They actively spread the quality message, making 23 quality leadership presentations to business, professional, and educational organizations in 1990 alone.

Zytec also shares its quality journey at its facilities. In 1990, representatives of 47 companies attended Zytec's "non-customer visit day" at its Redwood Falls manufacturing facility, where they learned what Zytec is doing to improve quality in areas such as Statistical Process Control, Just-In-Time manufacturing, Deming's 14 Points, and world-class manufacturing.

CATEGORY TWO: INFORMATION and ANALYSIS

In every department and at every level, Zytec is data driven. Zytec systematically collects and processes reliable, timely, and accurate data and information, of broad scope and significant depth.

The Long-Range Strategic Plan (LRSP) process establishes the key elements for quality improvement. Zytec selects data and information that support these elements. The company's data requirements are continually evaluated and improved through monthly MBP reviews, monthly senior staff meetings, weekly staff and departmental meetings, and ongoing functional and cross-functional task team meetings focused on specific quality objectives.

Zytec's most valuable data and information are customer-related and are used to plan product design and development and to continually improve product and service processes throughout the company. Examples of customer-related data are delivery time, warranty repair, customer satisfaction levels, and mean time between failure (MTBF).

Data on internal operations and processes help Zytec measure progress toward its Six Sigma goal. To that end, process measurements have been and are being converted to parts per million (PPM) on all product lines.

These and other data are collected and analyzed following standardized processes and techniques, and then used for proactive process improvements. The five major processes used to analyze quality data and information are competitive comparisons and benchmarking, MBP, internal audits, trends and surveys, and reliability analysis.

To improve the analysis, focus on eliminating root causes, and standardize the results, all Zytec employees have been trained in a seven-step problem-solving process called The Quality Control (QC) Story (Figure 19-1).

Zytec follows a six-step benchmarking process: plan, research, observe, analyze, adapt, improve—then start the cycle again. Each process or parameter to be benchmarked is assigned to a specific department, which assembles a team to identify world-class leaders for its process/parameter. The team chooses the data collection method, projects future performance levels, establishes goals, and develops and implements specific action plans for improve-

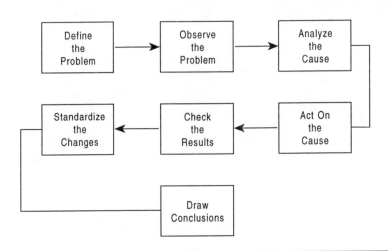

Figure 19-1 The QC Story

ment. In the past two years, Zytec has made benchmarking comparisons with several leading organizations, including Xerox, Motorola, Federal Express, IBM, DEC, Sony, 3M, Hewlett-Packard, and Kodak.

CATEGORY THREE: STRATEGIC QUALITY PLANNING

Zytec's planning process takes the quality and service needs of customers and drives them through the organization and to its suppliers. The process involves three steps:

- Data is gathered.
- Goals are set by long-range strategic planning cross-functional teams.
- Detailed action plans to implement these goals are developed by departmental MBP teams.

The Deming approach to setting goals and developing plans for quality leadership requires that planning is based on data. Zytec collects this data by soliciting customer feedback, conducting market research, and benchmarking customers, suppliers, competitors, and industry leaders.

Long-range strategic planning begins with small teams researching assigned focus areas. Each team prepares a report identifying what has been examined and how progress will be measured. Their reports are critiqued at two off-site, one-day planning meetings attended by all managers, all exempt employees, and representatives of the non-exempt and Multi-Functional Employee (MFE) work force (a total of more that 150 employees in 1990). At the end of these meetings, a document incorporating the presentations and discussions is written.

The document and briefings by members of the group are used to involve the rest of the work force in the LRSP process. When agreement is reached on the long-range plan, the focus shifts to short-term planning. At the same time, Zytec shares the LRSP with major customers and suppliers, who are asked for their reaction to the plan.

Using the LRSP and Zytec's corporate objectives as a road map, each operating department identifies what its MBP objectives are for the next fiscal year and how it will achieve them, developing detailed action plans with measurable, specific monthly goals. *In this way, Zytec's major strategies are converted into measurable goals.*

Finally, the short-term action plans are converted to financial plans. Detailed department budgets, manpower plans, and capital plans are established that support the objectives.

The CEO reviews each team's objectives and action plans. During the fiscal year, senior management reviews goals and progress toward those goals every month. Each department describes its progress in achieving its objectives, then reviews the action plan for one of its four objectives. Successes are commended. Problems are discussed and new actions or possible solutions proposed.

Because of the broad involvement and cross-functional development of the LRSP and MBP, Zytec's direction for the future, both short- and long-range, has broad consensus and support. It exemplifies Dr. Deming's 14th Point: "Put everybody in the company to work to accomplish the transformation. The transformation is everybody's job."

CATEGORY FOUR: HUMAN RESOURCE UTILIZATION

As the LRSP and MBP processes show, Zytec involves its human resources in setting and achieving the company's quality objectives.

Human resource planning is guided by Dr. Deming's 14 Points and the LRSP. The results of a recent planning cycle provide an example of how these are translated into short- and long-term objectives.

- **Deming Point 7:** Institute leadership. The aim of leadership should be to help people and machines and gadgets do a better job. Leadership of management is in need of overhaul, as well as leadership of production workers.

- **LRSP:** Implement self-managed work groups in which employees make most day-to-day decisions while management focuses on coaching and process improvement.

- **Long-Range Strategic Objective:** Managers will be trained to become better coaches/facilitators.

- **Short-Term Human Resources Objective:** Managers will become facilitators of self-managed work groups.

Several self-managed work groups are already in place. Employees in these groups, in cross-functional teams, and as individuals are granted broad authority to achieve their team and personal goals. For example, any employee can spend up to $1,000 to resolve a customer complaint without prior authority, hourly workers can make process changes with the agreement of only one other person, and sales people are authorized to travel whenever they feel it is necessary for customer service.

Zytec encourages employees to contribute to continuous improvement by providing opportunities to grow and participate and through extensive quality-oriented training.

Zytec's MFE system was designed by production workers for production workers. The identified tasks, when mastered, would make the employee more flexible and increase his/her pay. An employee group monitors and improves the MFE system.

All non-MFE employees develop personal action plans with their managers that focus on improved performance. This program, called FOCUS, was also developed by an employee team.

All Zytec employees have received quality training. As of February 1991, the average employee had received 72 hours of internal quality-related training. In 1990, every employee averaged 2.7 quality-related courses.

Mandatory quality training for all sales and marketing employees, engineers, technicians, managers, program writers, office clerical, MFEs, materials, accounting, personnel, and executives includes courses in Statistical Process Control, QC Story, Just-In-Time manufacturing, and "Service America." The training is provided by employees and by Zytec's full-time training department. New skills are reinforced by giving employees the authority and opportunity to use them. Ongoing assistance is available through in-line trainers in each production line, the training department, and managers at all levels and in all departments.

The result of training, involvement, and empowerment is that Zytec's employees believe Dr. Deming's 14 Points are more than vague guidelines. Since 1984, Zytec has surveyed employees seven times to gauge how effectively the company has implemented Deming's Points. As Figure 19-2 shows, Deming's Points provide an accurate description of the way Zytec conducts business.

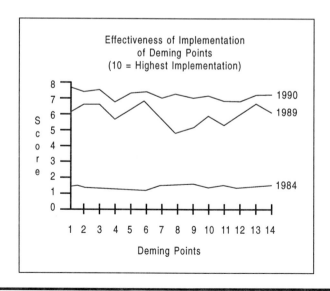

Figure 19-2 Implementation of Deming's 14 Points

CATEGORY FIVE:
QUALITY ASSURANCE of PRODUCTS and SERVICES

Zytec manufactures power supplies designed and manufactured to a customer's specifications or requirements (Black Box) or manufactured to a customer's design and documentation (White Box). Both follow a new product development process that includes four phase gate reviews by executive management.

Phase 1 is *Initiation.* A cross-functional team assigned to the project identifies customer quality needs through such means as Quality Function Deployment charting with the customer.

Phase 2 for the Black Box is *Design Initiation.* Various approaches, including the use of Zytec's Quick Custom® power supply circuits, can be used to meet customer needs while shortening the design-to-introduction cycle. For the White Box, Phase 2—*Project Documentation/Materials Planning*—involves qualifying components and suppliers.

Phase 3 for the White Box is *Sample Delivery and Customer Acceptance.* For the Black Box, Phase 3 is the *Prototype Delivery and Customer Verification Testing (CVT) Phase.* Engineers test prototype units to determine compliance with specifications. The tests include strife testing, which is stress testing beyond specification limits on thermal and power parameters, and CVT, which is detail testing to the specification limits. Feedback from the prototype assembly and testing is used to improve the design prior to production.

Phase 4 for both the Black Box and the White Box is *Preproduction and Process Certification.*

In addition to executive reviews at the end of every phase, reviews are held with customers and internal functions to ensure quality specifications are being met.

Zytec controls the processes used to produce its power supplies through an integrated Process Quality Control system.

■ New Product Introduction teams work upstream with design engineering and customers to assure that design requirements are met during manufacturing and testing.

■ Zytec's formal documentation system controls all engineering and manufacturing processes.

■ A process flow analysis is completed for every product, defining every process step together with its associated data collection and Statistical Process Control application. A process and/or procedure is developed and provided to manufacturing for each of these process steps.

■ Process capability assessments are conducted on all critical parameters during a product's certification process, allowing Zytec to measure the variation of these parameters from the product specification limit.

- Statistical Process Control provides another means of monitoring process variations for feedback and corrective action.

- Design of Experiments is applied to control, improve, and optimize processes.

- Monthly quality assurance audits help control processes through physical surveys.

Out-of-control occurrences and their root causes are identified by measuring process parameters and comparing the results to calculated variation limits. When an out-of-control occurrence is found, the root cause is analyzed and isolated and corrective action is taken.

Zytec involves all employees in improving its processes, products, and services by empowering them to identify and implement process improvements, by creating manufacturing focus teams, and by dedicating the company's value analysis department to assuring improvement in company-wide processes and supporting technical change throughout Zytec.

Zytec assesses quality, quality systems, and quality practices through a variety of assessment tools, including department system audits, product audits, quality climate surveys, Deming surveys, process capability analysis, strife and CVT testing, and MBP audits.

The quality improvement focus extends beyond engineering and manufacturing to every function within Zytec and to its suppliers.

Zytec's MBP process involves all employees in all departments in setting company goals and departmental objectives, and in developing action plans to meet them. By recognizing that each department produces a product or service and identifying the recipient as the customer for that product or service, everyone in Zytec has been able to identify ways to improve the quality they are delivering.

For Zytec's suppliers, the principal quality requirements, to be achieved by 1995, are: Six Sigma levels of quality, 96 percent on-time delivery to the day, and 25-day lead time. These three requirements are underscored in all communications with suppliers and supported by technical assistance and training.

Zytec uses three procedures to make sure its suppliers are meeting the company's quality requirements:

- **"Upstream" Supplier Selection and Communication Procedures.** Zytec is committed to open communication and the development of true partnerships with its suppliers. Zytec often involves suppliers in the early stages of a development program to benefit from their knowledge.

- **Receiving Inspection and Dock-to-Stock.** Critical parameters have been defined for each commodity so that Zytec can complete capability studies at incoming inspection. The capability study dictates the number of lots to be inspected before going dock-to-stock.

■ **Supplier Certification.** This program defines Zytec's quality requirements for its suppliers. At present, the program defines a PPM level for each commodity and a delivery percentage which is generic for all commodities.

CATEGORY SIX: QUALITY RESULTS

Based on extensive customer data and information, Zytec has defined, monitored, and improved the most relevant customer performance measures on an ongoing basis. Comparisons to competitors and to world leaders show that Zytec is above the industry average in all measures for which comparisons can be made, is the industry leader in more than half of the measures, and has the capability to become the world leader by executing its processes, strategies, and long-range strategic plans.

■ Product quality, as measured by the customer's out-of-box quality level, has improved from 99 percent in 1988 to 99.7 percent in 1990.

■ Zytec's product reliability, expressed as mean time between failure in hours, ranks in the world leader class when compared to like-complexity products. It has improved by a magnitude of ten in just five years.

■ On-time delivery is measured on a Statistical Process Control chart to promote continuous improvement. Process variance has been reduced at the same time that on-time delivery has improved.

■ Warranty costs were reduced by 48 percent from 1988 to 1990.

■ Zytec's ability to turn around a product sent to it for repair was reduced by 31 percent from 1988 to 1990.

■ Design cycle times were reduced by 51 percent from 1987 to 1990. Over the same period, safety approval cycle times were reduced by 50 percent and printed circuit board layout time was reduced by 69 percent.

■ From 1988 to 1990, product costs were cut by 30 to 40 percent, depending on the product line.

Zytec has seen similar improvements in the business processes, operations, and support services that make it possible for the company to meet its customers' quality requirements.

■ Manufacturing internal process yields are critical performance measures. From 1988 to 1990, Zytec improved internal yields by 51 percent to a level that is rapidly approaching Six Sigma.

■ Over the same period, Zytec reduced manufacturing cycle time by 26 percent. The improvement is 16 times the original plant performance.

▪ Inventory turn performance measures the effectiveness of inventory control. Zytec is at the world leadership level after improving inventory turns 128 percent from 1985 to 1990.

▪ Zytec's average monthly scrap rate has been cut in half since early 1988.

▪ Zytec focuses on three measures of improved profitability: increased use of standard parts (improved 333 percent from 1987–1990), sales/revenue per employee (compound growth rate of 15 percent from 1987–1990), and manufacturing direct labor as a percent of cost of sales (improved 57 percent from 1988–1990).

Zytec also monitors all key indicators of supplier performance, including PPM for all incoming material, percent on-time delivery, rejected materials, and percentage of lots received that are dock-to-stock. In every case the measurements have improved.

CATEGORY SEVEN: CUSTOMER SATISFACTION

Zytec knows its customers. It relies on 18 different processes to gather data and information from and about customers (Figure 19-3).

Many of these processes provide several sources of data. For example, the demographic surveys describe market size, growth rate, technology trends, and other characteristics Zytec uses to segment markets and define competitors and customers. In 1990, Zytec participated in eight different surveys that helped identify product and service quality features valued by customers.

Zytec is an industry leader at determining customer needs because it uses several processes to compile and validate the data it collects. The most valuable process is sharing results with customers and listening carefully to their perceptions of the findings.

Zytec knows its customers. Since the company started in 1984, every Zytec customer has been profiled. Of Zytec's 26 new build customers in that period of time, 20 were still customers in 1990. The company's major repair customer base has grown from 15 in 1988 to 26 in 1990.

Since most customers have changed from a warranty focus to an improved PPM or MTBF focus, Zytec's primary commitment to its customers is constant improvement in the features they value most: product quality (measured in PPM), product reliability (measured in MTBF), and on-time delivery.

Zytec believes all employees are customer-contact personnel. In 1990, over half of Zytec's employees had direct customer contact.

All employees receive customer relationship training. Specialized training is provided for sales representatives and account managers. These primary service providers are given the tools and technology they need to serve customers, and they are rewarded and recognized for their efforts. Zytec improves its role in customer relationships by listening to customers, surveying

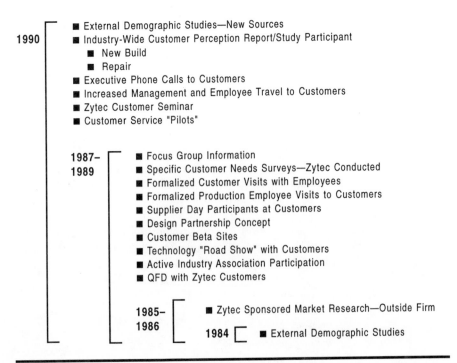

1990
- External Demographic Studies—New Sources
- Industry-Wide Customer Perception Report/Study Participant
  - New Build
  - Repair
- Executive Phone Calls to Customers
- Increased Management and Employee Travel to Customers
- Zytec Customer Seminar
- Customer Service "Pilots"

1987–
1989
- Focus Group Information
- Specific Customer Needs Surveys—Zytec Conducted
- Formalized Customer Visits with Employees
- Formalized Production Employee Visits to Customers
- Supplier Day Participants at Customers
- Design Partnership Concept
- Customer Beta Sites
- Technology "Road Show" with Customers
- Active Industry Association Participation
- QFD with Zytec Customers

1985–
1986
- Zytec Sponsored Market Research—Outside Firm

1984
- External Demographic Studies

Figure 19-3 The 1990 Menu of Zytec Processes to Determine Our Customers' Requirements and Expectations

customers, and benchmarking its sales/service practices against eleven world-class organizations.

Customer service personnel develop Zytec's customer service standards. Short- and long-term customer expectations are identified in the LRSP and deployed and tracked through the MBP process.

All complaints about Zytec's products or services are managed by the Vice President of Marketing and Sales. The complaints are formally measured, monitored, and resolved through Zytec's Customer Action Request process. Root causes are determined and the process improved to eliminate the cause.

Zytec relies on nine sources to determine customer satisfaction: sales representatives, account managers, quality reporting, executive phone calls, executive customer visits, employee customer visits, customer visits to Zytec, multi-client customer satisfaction surveys, and Zytec-specific customer satisfaction surveys. The data from these sources are reviewed and compared, and trends in customer satisfaction and changes in customer needs are tracked. This information is cycled through the LRSP process, then cascaded through annual plans, monthly plans, and daily actions.

Zytec's customer satisfaction trends show continuous improvement. For its new build product, Zytec's annual OEM revenue grew 74 percent from 1984 to 1990, compared to the industry annual growth rate of 8 to 10 percent. In an independent survey of power supply manufacturers, Zytec ranked #1 against its competitors and exceeded the industry average in 21 of 22 attributes deemed important to customers.

Satisfaction with Zytec repair also continues to grow, as reflected in its OEM repair revenue growth rate, which averaged 45.2 percent annually from 1988 to 1990.

Over the same period, adverse indicators have declined. For example, total warranty costs as a percent of revenue dropped 36 percent.

Since 1987, Zytec has been recognized by its customers with nine awards for supplier excellence. Zytec remains the sole source supplier for 18 of its 20 new build customers.

Zytec's growth is a direct result of its commitment to providing its customers with quality products and services, a commitment captured in the three words chosen to guide the company in 1984: *Quality, Service,* and *Value.*

20

PATIENT DISCHARGE/TRANSFER: A TQM CASE STUDY

INTRODUCTION

There is an increasing need to balance the demands of reform-minded health care consumers and the need for cost containment. The complexities created by these needs have become the major driving force in the decision by health care organizations to embrace the philosophy of Continuous Quality Improvement (CQI). While the concept of CQI (or Total Quality Management [TQM] as it is known in the manufacturing sector) has been in existence for many decades, health care organizations have been slow to adopt it. CQI is a leadership philosophy driven by the following chain reaction: quality improvement results in decreased costs because of fewer "defects," less rework, fewer delays, better use of resources, increased productivity, better quality, and lower price. All of these facilitate capturing the market, staying in business, and providing more jobs. Although CQI is not a panacea for all the challenges facing health care organizations, it holds significant promise for the health care industry, particularly for organizations that seek to dominate the market. The following case study describes one hospital's attempt at using CQI to gain a competitive edge.

Reprinted from Aristides Pallin, Vincent K. Omachonu, and Seema Prashad, "Patient Discharge/Transfer: A TQM Case Study," *Quality Engineering,* Vol. 7, No. 2, 1994 by courtesy of Marcel Dekker, Inc., New York.

CASE STUDY

The study reported in this case is based on the application of the concept of CQI at a local 500-bed licensed acute-care facility in South Florida. The facility is located on a campus which houses a stand-alone MRI, an out-patient center, and a professional office building. The hospital offers a wide range of services, including diagnostic testing, a full-service operating room (which includes open heart surgery), obstetrics, and an emergency room. The hospital's current census is at an average of 290 patients per day, with a total FTE count of approximately 1500 employees. Although licensed for over 500 beds, the hospital currently operates at about sixty percent of that capacity.

Increased competition and a reduction in reimbursement have forced the hospital into a state of fiscal restraint. The increase in the portion of payers who pay on a prospective basis has been largely responsible for the reduction in net revenues. Such payers include Managed Care, Medicare, and Medicaid. Top management is committed to the concept of CQI as a way of achieving the goals of cost reduction through the adoption of a never-ending process of quality improvement.

PROBLEM

The overall census is increasing at the hospital, especially in the intensive care areas, leaving fewer beds available for patients entering the hospital. As a result, the process of discharging or transferring a patient and preparing that room for a new patient has become critical to hospital operations. Slow turnaround time can result in extended stays in the emergency room, recovery room, or admitting, which in turn leads to patient, physician, and family member dissatisfaction. Several incidents have prompted administration to request a detailed analysis to be conducted by the Management Services Department. Management Services is the department responsible for management engineering activities at the hospital.

The process of discharging or transferring patients is complex and involves several departments. For this reason it was decided to use a team approach to solving the problem. In selecting the team members, all departments that are involved in the process were invited to participate.

To address this issue, the following problem statement was developed and presented to those invited to attend the meeting:

> ▪ Evaluate and improve the overall process of discharging or transferring patients from the time the order is written to the time the room is ready for a new patient to better meet the requirements of our customers.

This statement is more an action statement or goal than a problem definition. This was done to assure that the process was viewed as a positive step to improve the service provided to the patient and not one of assigning blame. The need to meet customer expectations is also identified to keep the team focused on a common goal. This definition was not easily accepted because of the prejudices many team members brought to the meeting. Suggestions such as "make the environmental worker clean faster" or "reduce patient escort response times and we can all go home" were discussed at the meeting.

TEAM MEMBERSHIP

A cross-functional team was developed with members from all the departments involved in the process. Department heads were invited and encouraged to send someone directly involved with the process to represent their department. The final team membership included the following personnel:

1. Head Nurse, Med. Surgical Floor
2. Head Nurse, Telemetry Floor
3. Manager, Admitting
4. Manager, Environmental Services
5. Supervisor, Patient Escort
6. Manager, Management Services
7. Management Engineer

The team consisted entirely of management personnel except for the management engineer. Some department managers were at times defensive and took every comment made by the members of the team personally. The fact that the team consisted of only management personnel was not a problem in any way. They were selected based on their expertise and ownership of the problem. A team leader was appointed, who worked hard to maintain the team's focus and guide it to its goals. The team leader was the Head Nurse, Med. Surgical Floor.

CUSTOMER DEFINITION

Training was provided on an as-needed basis and was given simultaneously along with the team activities. In the initial meeting the definition of customer was evaluated by each team member. This was an interesting process as each member soon discovered that there are both internal and external customers. The expectations of each customer were evaluated in terms of the process, leading to the following definitions:

External customers	Expectations
Patient	Timely and smooth admission and transfer process
Family member	Same
Physician	Same

Internal customers	Expectations
Nursing	Quick response from patient escort
Environmental	Notification that the room is empty
Patient Escort	Ready patients when arriving on the floor
Admitting	Accurate room status in computer system

The above exercise revealed to the group members that each one of them was dependent on the others for optimal performance. This realization marked the beginning of the team-building effort. The underlying customer was always considered to be the patient. Patient convenience would translate into good will and satisfaction of the patient and other external customers.

PROCESS ANALYSIS: CQI TOOLS

This section will closely follow the analysis conducted by the group. It was the responsibility of the Management Services personnel to monitor the study. The approach taken was to use as many CQI tools as deemed applicable and necessary to reach an improvement of the process. Some of the tools that will be described were not fully explored by the team, because of time constraints and the lack of previous experience by team members. These tools were developed by Management Services and later shared with the team. Also, simplified versions of the tools were utilized.

Flowchart

This is the first tool presented to the team and in many ways the most valuable. The flowchart is a graphic representation of the flow of a process. The group members were trained with two basic building blocks of a flowchart. These blocks were the operation and decision. The chart was developed by the group as a whole, with individual departments producing charts of their own operations. The complexity of the process required several meetings and observations before the final flowchart was approved by the group.

The process begins in one of two possible ways. The first is when a physician orders a discharge or transfer for the patient. The other situation is when a nurse requests a transfer because of nurse or patient preference. Only the physician can order a discharge or transfer for medical reasons. Once this is completed, Patient Escort is usually called and moves the patient. Not all moves are made by Patient Escort, however. When the patient leaves the room, the status in the hospital ADT (Admission, Discharge, Transfer) system is updated. This update could be made by several individuals, depending on where the patient went and who moved the patient. At this time the Environmental Services worker is responsible for checking the "status board" to see if the patient is out of the room. This process requires the employee to walk to the board continuously for updates. When the employee realizes the room is empty, the cleaning begins. Upon completion of the cleaning process, the supervisor in Environmental Services is notified and in turn updates the ADT system to reflect a "clean" status. This allows Admitting to place a new patient in the room.

The resulting overall chart of this process is shown in Figure 20-1. Evaluation of the flowchart revealed two major areas of concern. The first is the lack of a centralized responsibility for entering "patient out of the room" status in the computer system. This is done by three different departments, depending on the patient. The second is the process by which the environmental workers determine if the room is empty and ready to be cleaned. The current process has the employee going to the manual status board or to the patient's room. This leads to unnecessary travel and the potential for having the room empty without the worker knowing it is ready for cleaning.

Data Collection and Analysis

One key element in all TQM efforts is the "management by fact" philosophy. This represents collecting data to quantify the current process, verify assumptions, and later determine effects of changes to the process. After the flowchart was completed, several recommendations were discussed but not implemented until more detailed information could be collected.

The data that were to be collected were geared to cover all aspects of the process. This was an immense task because of the number of observations and data elements. For this reason, ten consecutive days were studied excluding weekends. Weekends were excluded because of the low consensus and the availability of rooms. Luckily, much of the data was available through the hospital's computer system. The following data elements were collected:

1. Type of process (discharge, transfer, expiration)
    (*Note:* For the remainder of the report discharges, transfers, and expiration will be called "events.")

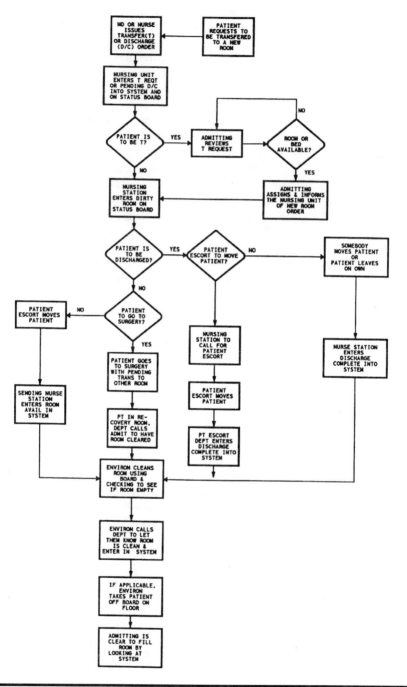

Figure 20-1 Flowchart for Patient Room Processing Post Discharge or Transfer

2. Room the event originated from
3. Room patient is transferred into (if a transfer)
4. Census on days studied
5. Time elements:
   a. Escort called
   b. Escort arrives on floor
   c. Escort leaves floor
   d. Discharge or transfer completed in system
   e. Cleaning of room starts
   f. Cleaning of room is completed
   g. Room is available in system

This information was then compiled and summarized to determine the key statistic of the process. Some of the more valuable statistics included the length of each component of the process. Also, information on the volume and time of day events occurred was tabulated. This data was then used to develop the TQM tools that will be discussed in the following sections.

Histogram

This tool is used to illustrate an observed activity versus a continuous scale, such as time. This will allow changes in activity over time to be reviewed in a bar graph format. Three histograms were developed to assist in the analysis of the process. These included the following:

1. Average transfers by time of day (Figure 20-2)
2. Average discharges by time of day (Figure 20-3)
3. Average discharges and transfers by time of day (Figure 20-4)

Hospital policy specifies that all discharges and elective transfers are to be completed by 11:00 a.m. For this reason the histogram's first bar depicts the midnight to 11:00 a.m. time frame. The histogram clearly shows this is rarely being adhered to. Also noted is the substantial number of events that are occurring after 5:00 p.m., when both Patient Escort and Environmental Services are working with skeleton crews.

Run Chart

A run chart is a line graph that shows trends of an indicator time. Run charts developed depict the events and census across each day of the data collection process. This tool resulted in a more concise realization of the volume and variability of the activity. These charts are illustrated in Figure 20-5 and Figure 20-6.

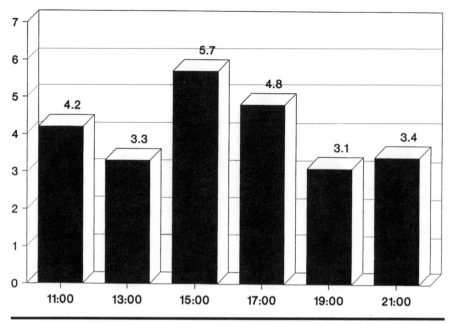

Figure 20-2 Histogram for Average Transfers by Time of Day

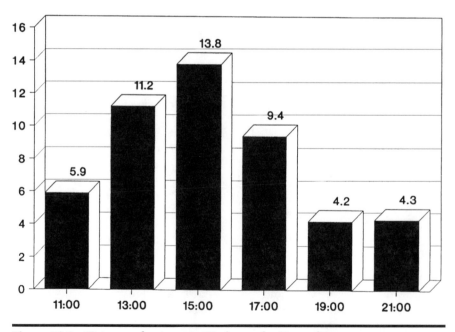

Figure 20-3 Histogram for Average Discharges by Time of Day

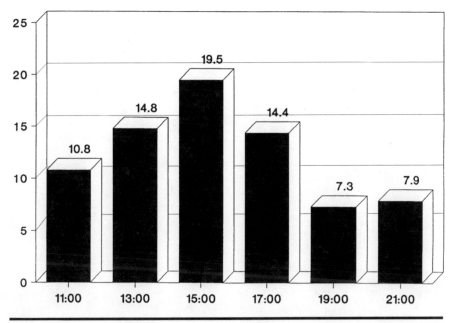

Figure 20-4 Histogram for Average Discharges and Transfers by Time of Day

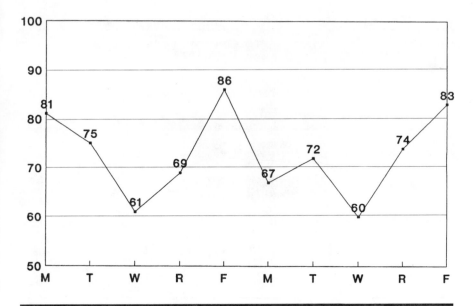

Figure 20-5 Run Chart for Discharges, Transfers, and Expirations

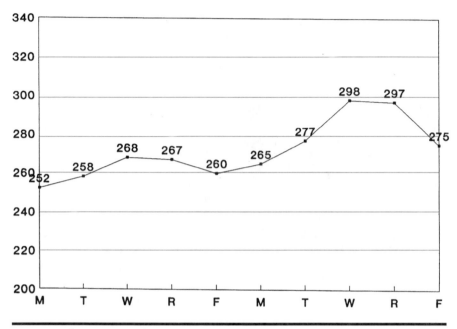

Figure 20-6 Run Chart for Census

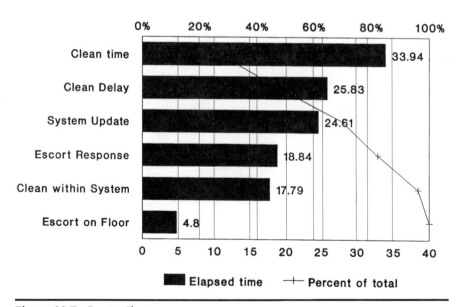

Figure 20-7 Pareto Chart

Pareto Chart

This chart takes the indicators collected, sorts them in descending order, and then calculates a cumulative percentage for each. Usually this is used to focus on the problem which is occurring most often. In our study, a variation of this chart was used to determine the components of the process which were the lengthiest. The chart in Figure 20-7 was developed using the mean duration of each component of the process. (*Note:* The actual chart has variations from the traditional Pareto chart due to limitations in the graphics software used.) The chart illustrates that the process of cleaning the room is the longest. However, three out of the next four components are actually delay time where no operations are taking place. If these could be eliminated or reduced, the overall process time could be greatly impacted.

Cause-and-Effect Diagram

This diagram is used to graphically represent the cause of problems in a process. The problems are usually broken down into people, machine, method, and materials. In our case we are looking for causes of a delay in the process. The resulting chart points to areas that can improve the overall operations and eliminate cases that have problems leading to extended turnaround times. The diagram is illustrated in Figure 20-8.

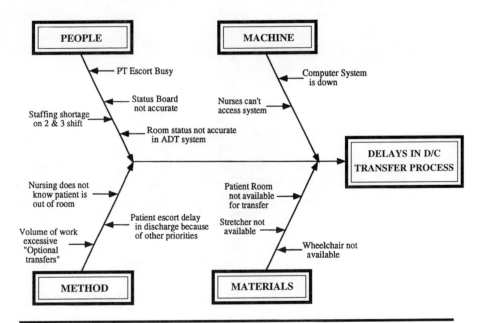

Figure 20-8   Cause-and-Effect Diagram

RECOMMENDATIONS

This process has a variety of issues, of which only a few will be addressed. The results of the changes will then be evaluated and further study will be conducted. A continuous process improvement will result. Initial recommendations include the following:

1. Issue all environmental workers beepers to eliminate the manual process of checking the status board and patient room. Supervisors will then beep the employees and inform them as soon as the room is empty and entered into the computer system.

2. Place responsibility for updating the room status in the system on the nurse manager. This will assign this responsibility to one person who will be accountable and eliminate the current ambiguity.

3. Allow patients with assistance from family members or friends and physician approval the option of leaving on their own. This will reduce or eliminate the waiting time for patient escort. With the other tasks patient escort performs, transfers and discharges are often of the lowest priority.

4. Research the possibility of stricter transfer policies, especially for those done within the same floor. This may reduce the total number of transfers and therefore allow the staff to more adequately meet the workload. Patient education would be necessary in order for the patients to understand how this would ultimately enhance patient satisfaction.

5. Change the current staffing pattern to eliminate the skeleton crews in the evening. This non-traditional staffing pattern should allow the staff to be more reflective of the actual workload.

CONCLUSION

The above recommendations may have been immediately obvious, but the benefit of the TQM process is also in ownership and participation. All the team members were able to constructively focus their efforts on improving a process without the need to assign blame to an individual or department. This is no small accomplishment.

Finally, the benefits of an analysis focusing on data and using TQM tools resulted in feasible recommendations that did not add staff or require much money. The results were a better product for the patient, improved employee morale through participation, and greater teamwork. In this exercise the TQM approach was a resounding success.

BIBLIOGRAPHY

Omachonu, V.K., *Total Quality and Productivity Management in Health Care Organizations*. Milwaukee, WI: Quality Press–American Society for Quality Control (ASQC), and Norcross, GA: Industrial Engineering and Management Press–Institute of Industrial Engineers (IIE), 1991.

21

LIBERTY BANK

OVERVIEW

Liberty Bank, a full-service bank that provides a complete line of financial and asset management services to consumer (retail), business (commercial), trust, and credit card customers, has been a fixture in the Baltimore, Maryland, standard metropolitan statistical area (SMSA) since 1886. Chartered in the State of Maryland, we operate 21 branches, including our headquarters branch in Baltimore's central financial district. We also own or operate 146 automatic teller machines (ATMs) throughout Maryland and participate in the CIRRUS and PLUS networks for transactions outside the state. Our competitors include approximately 100 commercial banks and 200 thrifts and savings banks chartered to conduct business in the State of Maryland; about half of these institutions actively compete in the Baltimore SMSA.

The quality initiative at Liberty Bank began in 1986 both as a response to increasing competitive pressure in the deregulated financial services environment and as a renewed commitment to customer satisfaction as we entered our second century of service to the community. We seek to differentiate ourselves from our competitors by delighting our customers with the products and services they receive, and thereby setting the standard not only for Baltimore-area banks, but for banks of comparable size nationwide. We also pride ourselves in the level of involvement that our customers have—through their participation in customers surveys, focus groups, and pilot introductions—in

Reproduced with permission. Prepared for use in the 1992 Malcolm Baldrige National Quality Award Examiner Preparation Course. The Liberty Bank II Case Study describes a fictitious company. There is no connection between the Liberty Bank II Case Study and any company, either named Liberty Bank or otherwise.

the development of new products and services. Bank teams also work with customers to determine how they use each product or service and how we can improve our products and services using Liberty's Integrated Process Management Approach.

Consumer and business banking and trust services are Liberty Bank's principal product lines. Major consumer banking products include checking and savings (including money market) accounts, certificates of deposit, credit cards, consumer loans, and first and second home mortgages. Our primary business customers include companies with annual revenues between $10 million and $100 million. Our services to this clientele include business checking and payroll accounts, cash management and remittance banking facilities, employee benefit and pension accounting services, and working capital lending and project financing. Trust services are available to both business and high-worth consumers and include overall financial planning and management, estate settlements, and Uniform Gifts/Transfers to Minors accounts. In recent years, we have grown our credit card operations, capitalizing on our historically low interest rates and annual charges to attract consumers from around the country. Because these operations are chartered in Delaware, rather than Maryland, they are treated as a separate line of business.

Across all customer groups, Liberty Bank has determined through extensive ongoing research that customers value eight service features above all others. All of our customers demand *timeliness* in distribution of reports and statements and responses to inquiries, uncompromising *accuracy, immediate transaction service,* and *knowledgeable, courteous* banking staff. Consumers also place a high value on *convenience* of location and business hours, while business customers add *accessibility* of account managers to their key requirements. Finally, business and trust customers place a high premium on the *longevity* of their account officers or teams; these customers do not wish to deal with a steady rotation of Bank officers and come to value the knowledge of their business that competent officers develop over time.

Customer research has also shown that the level of performance that customers expect of each of these eight service features varies by type of service encounter. For example, timeliness in posting a deposit is quite different from timeliness in mailing a credit card statement. Accordingly, since 1989, we have tracked our performance for each service feature by each of the 24 specific service encounters we have identified. These encounters are then rolled up into six categories that represent the most important service encounters identified by customers. These categories, the service features that characterize them, and the customer groups affected by the relationships are shown in Table 21-1.

Within the financial services industry, differentiation among banks and thrifts is largely based on price (interest rates and service charges), convenience (of locations and hours), and service (accuracy and responsiveness). National and regional surveys have consistently shown that customers do not

Table 21-1 Service Encounters and Features

Account feature	Account opening	Transactions	Account changes	Account statements	Relationship management	Inquiries
Timeliness	C, B, T, CC	C, B, T	C, B, T, CC	C, B, T, CC	B, T	C, B, T, CC
Accuracy	C, B, T, CC	C, B, T	C, B, T, CC	C, B, T, CC	B, T	C, B, T, CC
Immediate transaction	C, B, T, CC	C, B, T	C, B, T, CC		B, T	C, B, T, CC
Knowledge	C, B, T	C, B, T	C, B, T		B, T	C, B, T, CC
Courtesy	C, B, T,	C, B, T	C, B, T		B, T	C, B, T, CC
Convenience	C, B, T, CC	C, B, T	C, B, T, CC	C, B, T, CC	B, T	C, B, T, CC
Accessibility	C, B, T	C, B, T	C, B, T		B, T	C, B, T, CC
Longevity					B, T	C, B, T, CC

Key: C = Consumer; B = Business; T = Trust; CC = Credit card.

feel that there are significant differences in the service levels offered by banks; they have therefore tended to select banks based on price and location convenience. This has given rise to a common industry saying: "You win them on price, but you lose them on service."

Rather than simply accept this widely held notion, Liberty Bank has worked hard to maintain consistently high service quality, and as a result enjoys a very strong competitive position in its market: It has captured approximately 17 percent of the total financial services market in the Baltimore area. More important, however, is its share of "primary relationships," those customers who use Liberty as their main bank; 75 percent of Liberty Bank's customers use it as their primary bank, a rate twice that of Liberty's competitors. Clearly, Liberty Bank's customers consider it to be a one-stop financial management center.

Liberty Bank's base of approximately 750 fulltime-equivalent employees is representative of the demographic and socioeconomic composition of the labor market and population within the greater Baltimore SMSA. More than 95 percent of our employees have received quality training, and the majority participate in problem-solving or quality improvement teams. Employees understand that they have the authority to do what is necessary to meet the needs of customers, more than 90 percent are very satisfied with their jobs, and turnover, at 8.7 percent, is less than half the industry average as well as the Baltimore-area average.

Liberty Bank also encourages supplier participation in the quality effort. Supplier certification programs ensure that all external providers of goods and services meet our high quality standards and deliver increasingly higher levels of service. Major suppliers include the Federal Reserve Bank System, which is operated by the U.S. Treasury and controls the national check-clearing system; the nationally franchised charge card networks; ATM processing networks; providers of such capital equipment as teller and ATM terminals, computers, and paper; and communication and telephone companies.

The success of our quality efforts is evident. We hold $2.1 billion in assets and have a very strong financial performance record. For 1991, our 2.1 percent return on assets and 20.4 percent return on equity were both at the 95th percentile for all banks in our nationwide peer group ($1 billion to $5 billion in assets). Over the past three years, quality improvement projects have yielded more than $2.5 million annually in increased profits. Our customer satisfaction levels have improved significantly and exceed those of our major competitors.

2.0 INFORMATION and ANALYSIS

2.1a Because Liberty Bank is a process-driven organization, it has determined that measurements are most meaningful and actionable when they are taken on a process basis. Each process at the Bank can be traced to a customer interaction, or *moment of truth,* whether external or internal. Therefore, the Bank makes its measurements and aggregates its data with respect to moment of truth (Table 21-2).

Table 21-2 Moments of Truth

Transactions with branch tellers	Telephone inquiries to Customer Information Center (automated attendant)	Account opening, closing, or changing
Transaction with drive-up window tellers	Telephone inquiries to Customer Information Center ("live" representative)	Product/service information requests from teller
Transactions with branch account representative	Telephone inquiries to branch	Product/service information requests from account representative
Transactions with brokerage representatives	Telephone inquiries (business account representative)	Electronic transfers
ATM transactions in the Bank	Telephone inquiries (trust account representative)	Other account-related Bank correspondence
ATM transactions at standalone kiosks or in stores	Written requests and complaints	Telephone banking
Loan applications	Statement receipt	Home banking
Loan inquiries	Bill payment	Visits from Bank management

Furthermore, each process at the Bank can be measured with respect to the service features that have historically been most important and most meaningful to customers. One or more of these features—timeliness, accuracy, immediate transaction service, knowledge, courtesy, convenience, accessibility, and longevity of the account officer/team—can be tracked for each moment of truth. Liberty Bank monitors its performance on 160 of these combinations of moment of truth and service feature, which encompasses nearly every transaction that occurs at the Bank. The quality and operational performance results for these 160 measures are stored and tracked in Liberty Bank's state-of-the-art Quality Management Information and Control System (QMICS). Developed in 1986 and refined annually since, QMICS is the heart of our fact-based quality management system, is accessible to virtually every employee through terminals at every workstation, and provides the data needed for companywide analysis (Item 2.3), process control and improvement, and performance monitoring.

The types and sources of data gathered by Liberty Bank are described in Table 21-3.

Table 21-3 Liberty Bank Quality Data

Source	Description	Quality planning	Bench-marking	Quality comparisons	Customer requirements
External					
American Banking Association	Nationwide data for commercial banks	✔	✔	✔	
Bank Administration Institute	National customer satisfaction norms	✔	✔	✔	✔
Bank Earnings Improvement Group	Reports on private banking studies	✔			
Bureau of National Affairs	Industry trends in human resources			✔	
Cole Associates	Data on employee turnover in financial firms			✔	
Company-Specific Benchmarks	Comparative data on processes and results	✔	✔	✔	
Decision Research Sciences	Deposit data for all financial institutions	✔	✔	✔	
Dun & Bradstreet	Market data and background reports on area companies	✔			✔
George Fredericks, Inc.	Reports on branch studies and distribution plans	✔	✔	✔	

Table 21-3 (continued) Liberty Bank Quality Data

Source	Description	Quality planning	Bench-marking	Quality comparisons	Customer requirements
			Uses		
Global Research Corporation	Market share and customer perception data	✔		✔	✔
Greenwich Associates	Surveys of business customer needs	✔	✔	✔	✔
Human Resource Management Association	Reports on HR trends and innovations	✔	✔		
Metromail, Atlas, Donnelly	Census data	✔			
Noah MCIF	Online customer data customized and segmented by Liberty Bank customers	✔			✔
Sheshunoff Associates Call Report	Data for commercial banks and S&Ls	✔	✔	✔	
Systems Administrators Association	MIS data, innovations, and trends		✔	✔	
Total Research Corporation Associates	Baltimore market surveys	✔	✔	✔	✔
TrustCompare	Peer-group comparisons using trust data	✔	✔	✔	
Internal					
ATMs	Data on ATM uptime and usage	✔			✔
Call Map Tracking System	Customer needs as expressed to call officers	✔			✔
Community Reinvestment Act	Information on community needs				✔
Complaint Tracking System (CTS)	Database of customer complaints	✔			✔
Customer Surveys	Statement inserts, direct mail, telephone surveys, focus groups, branch customer interviews	✔	✔	✔	✔
Employee Survey	Annual survey of all employees	✔			
Financial Management and Information Control System (MICS)	Comprehensive online financial reporting system	✔			
Quality Management and Information Control System (QMICS)	Comprehensive online operational performance reporting system	✔	✔	✔	

2.2a The Bank uses five different groups for competitive comparisons and benchmarks:

- *World class.* Those companies that, regardless of their industry, perform a function (e.g., billing) or business process (e.g., budgeting) better than any other company.
- *Top banks.* A national consortium of 17 high-performing banks (including Liberty) with $1 billion to $5 billion in assets.
- *Top Baltimore banks.* Liberty's two largest competitors in the Baltimore area.
- *Baltimore industry average.* Liberty's direct competitors in the Baltimore market, including the two in the above category.
- *Industry average.* National data for statistics available only on an industry-wide basis, e.g., employee turnover and loan loss experience.

Liberty uses customer surveys and data from bank associations and functional associations (e.g., the Association of Market Researchers) to identify the areas most critical to its success and to decide which practices of which comparison groups to benchmark.

2.2b The sources of competitive and benchmark data are shown in Item 2.1a.

2.3 Analysis and Uses of Company-Level Data

2.3a Liberty Bank's primary data- and information-analysis approach is a structured problem-solving approach called IDEAS. Work groups and quality improvement teams (QITS) throughout Liberty Bank use IDEAS to prioritize improvement opportunities, perform root-cause analysis, develop and implement improvements, and standardize and measure results. IDEAS comprises five broad steps:

- **I**dentification and selection of a project or problem involves such techniques as quality measure review, quality cost identification, flowcharting, work process documentation, sampling and surveying, brainstorming, Pareto analysis, and customer satisfaction research.
- **D**ata collection includes direct observation, interviewing, sampling and surveying, statistics development, market research, and company and competitor shopping.
- **E**valuation of data (to determine a solution) is done by Pareto analysis, work simplification, cause-and-effect analysis, brainstorming, trend analysis, segmentation analysis, benchmarking, buyer behavior analysis, and cross-tab analysis.

▪ **A**ction taken based on the evaluation includes pilot programs, scenario tests, simulation programs, reports of projected benefit-to-cost ratios, and forecasts of quality improvements resulting from changes in practice.

▪ **S**tandardization and ongoing measurement of performance results is effected by documentation of improvement actions and results, communication of improvements through meetings, newsletters, etc., and application of statistical measurements using such tools as run charts, control charts, Pareto analysis, and sampling to quantify results.

We use IDEAS in two ways: to solve specific problems at all levels and as an integral part of our Integrated Process Management (IPM) approach, as described in Items 5.2 and 5.3. In analyzing customer-related data at a Bankwide level, we aggregate the data not only by customer segment and by service feature, but also by type of service encounter, or *moment of truth*. Thus, for example, Liberty Bank aggregates all data related to personal transactions that occur *between a customer and a teller inside a branch* (to be distinct from such transactions between a customer and an account representative). As described in Item 2.1, we have determined that there are 24 different moments of truth that are most important to our customers.

Once the customer-related data have been identified and collected (the first two steps in IDEAS), they are then analyzed, again using IDEAS. This involves evaluating the data for possible shortfalls and areas for improvement. This evaluation is done first by moment of truth and then aggregated *across* moments of truth to reveal problems inherent in common work processes. Thus, low customer satisfaction ratings for timeliness of response to both business and trust inquiries, for example, may imply one of two things:

▪ A need to improve the timeliness of updates to QMICS, the Bank's online information system

▪ Inadequate training of business and trust representatives

Further analysis, also done within the evaluation step of IDEAS, would reveal the true root cause. IDEAS is also a useful tool for prioritizing solutions to Liberty Bank's customer-related problems.

Aggregation of customer-related data by moment of truth also enables us to correlate our performance on internal measures with customer satisfaction and retention. Customers can readily relate the satisfaction they feel with specific instances of good or bad service. Lost customers, whom Liberty Bank interviews regularly, are especially able to point to negative experiences at moments of truth as reasons for leaving (that is, if they did not leave for reasons of location). In particular, detailed analysis of former customers' reasons for leaving enables the Bank to focus on aspects of its performance that are more likely than others to cause customer dissatisfaction. Thus, when the Bank understood that its inability to resolve problems quickly and to the customer's satisfaction was the primary cause of defections (see Item 7.4), it was able to address the problem, determine root causes, and develop effective solutions.

2.3b Liberty Bank's operational performance data are also aggregated by customer moments of truth. Thus, key internal measures described in Item 6.1, which include timeliness, accuracy, immediate transaction service, and knowledge, are tracked (where they are relevant) for each moment of truth. Again, IDEAS and the process management tools described in Item 5.2 help Bank personnel pinpoint areas for improvement and develop solutions for them.

As an example of the usefulness of the Bank's improvement processes, teams of telephone representatives were able to reduce such sources of customer dissatisfaction as frequent busy signals, blocked calls, long holds, and abandoned calls. Pareto analysis of incoming-call volume by time of day revealed that customers tended to call the Bank during three time periods: 8:30–9:30 A.M., 11:30 A.M.–1:00 P.M., and 4:00–5:00 P.M. The teams recommended that the Bank include with monthly statements a notice of "preferred" calling times and further lobbied management for increased staffing during the peak hours. Within two months, the number of busy signals fell from more than 1200 to fewer than 80 per month and average hold time dropped from 3 minutes to less than 30 seconds. Perhaps more impressive is the fact that these results were obtained *before* the Bank installed its automated call-answering and -routing system and 24-hour Customer Information Center.

2.3c Through data correlation and analysis, Liberty Bank has determined that certain levels of customer satisfaction can reliably predict customer retention and market-share growth. We have also found that satisfaction levels lower than 80 percent for certain combinations of moment of truth and service feature can be directly mapped—after accounting for time lags—to losses of customers. Accordingly, we pay special attention to trends in these data.

2.3d The on-line availability (in QMICS) of data permits employees and teams to evaluate the effectiveness of different methods of data collection. Because QMICS also stores reports and process improvement data from QIT activities, teams embarking on similar initiatives can learn from the successes (and mistakes) of others. Not only does this prevent duplication of work, but it also enables the Bank to improve team effectiveness.

Through the IDEAS process we have developed automated data analysis techniques and standardized data files and sources in QMICS. We encourage employees to use the process not only as members of quality improvement teams, but also to improve the quality of their daily work. To reinforce this, we use actual case studies of improvement efforts in our weekly training sessions to increase the proficiency of employees in using the process.

Data gathering and analysis are reviewed by management and employees, individually and in teams, to assess their effectiveness in providing the information necessary to manage the business and meet the Bank's quality objectives. These reviews frequently identify new types of data to collect and ways to improve the predictive value of data such as replacing one data element with a more predictive element. Finally, this analysis also identifies opportunities to improve the analytical techniques to be applied to the data.

5.0 MANAGEMENT of PROCESS QUALITY

5.2 Process Management— Product and Service Production and Delivery Processes

5.2a As a process-driven institution, Liberty Bank is well equipped to design and maintain its production and service delivery processes. As described in Item 5.1, the design of each product, service, and delivery process includes the development of a rigorous map that identifies each process and includes flowcharts and process descriptions. When a new process is designed or an existing process is refined, both in-process and end-of-process measurements and performance targets and limits are established. Each employee performing a step in the process receives complete process documentation, which includes the process map as well as maps of other processes that rely on or support their process. This documentation is used extensively for employee training and the establishment of performance standards for all process steps. Performance to each standard is tracked by the QMICS system and reviewed daily by employees, weekly by line and staff managers, and monthly by the Senior Management Quality Council (SMQC).

We track two types of key quality measures for each of our key processes:

■ Those related to customer moments of truth and the major service features—timeliness, accuracy, immediate transaction service, knowledge, courtesy, convenience, accessibility, and longevity—that determine customer satisfaction at each interaction. Our performance for some of these features is recorded using QMICS; for example, timeliness in answering telephone calls is electronically recorded each time the receiver is picked up, and call abandonment (one element of accessibility) is likewise recorded electronically. Our performance for other features, such as courtesy and knowledge, is assessed through direct observation and customer surveys and then input to QMICS.

■ Internal (efficiency) measures that may not be visible to the external customer. Liberty Bank tracks process cost, rework, and cycle time to ensure that its operations are both efficient and effective. Again, these data are then input to QMICS.

Liberty Bank also conducts routine operational audits as part of its management process. Line managers and quality teams periodically sample actual transactions or batches of work to assure product and service quality and to verify that the process is being performed as described in the process documentation.

All Liberty Bank product and service production processes are designed to detect, correct, and prevent process upsets. Whenever QMICS reports, operational audits, or random process sampling indicates a real or potential process upset, the appropriate process improvement team investigates. Simple process upsets caused by easily identified lapses in procedure or communication are usually resolved immediately. For more complex upsets, the process improvement team uses the IDEAS process to identify the root cause of the problem and recommend corrective steps. As described in Item 2.3, IDEAS provides several statistical tools for analysis of the data the team collects. These include fishbone diagrams, force-field analysis, Pareto charts, and process maps (to determine where in the process the upsets are occurring).

If the root-cause analysis reveals that the process was functioning properly, but that certain inputs or work tasks were below par, then the corrective action usually requires a reinforcement and reaffirmation of the importance of controlling the process. The team monitors the process for one week, to ensure that it is functioning as expected. Sometimes, of course, the team's analysis shows that the process upset arose from a basic flaw in the process; perhaps the work load has overstressed the original design of the process, or perhaps a series of subtle (and perhaps unnoticed) changes in customer requirements has rendered the process inefficient. When this is the case, the team recommends that the process be reconsidered and perhaps redesigned. The approach for doing so is described below.

5.2b Because processes operate horizontally through an organization and often involve several vertically managed departments or units, some companies have experienced difficulties in improving their processes. Liberty Bank has overcome this obstacle through the creation of *process champions,* managers who are responsible for constantly overseeing the performance and leading the ongoing improvement of their assigned processes. Each of Liberty Bank's 60 processes has a process champion, who works with a cross-functional process improvement team (PIT) that comprises employees and/or managers from each department participating in the process.

During our annual goal-setting and planning effort, the process champions and their PITs establish specific process improvement objectives for the process and each process participant. Often, the objective is to reduce process variability and continue to meet customer expectations. When the process has met or exceeded expectations for at least four quarters, however, it may need to be "challenged"; we believe that there is always a way to "make it better."

In other cases, the objective may be to improve process performance at "process intersections," those steps in each process that rely on outputs from or provide input to other processes. Effective communications *between* process champions is therefore critical to the effective performance of each process.

Process improvement is generally performed through the Bank's integrated process management (IPM) methodology. This three-phase, eight-step approach, which is a variation of the common plan-do-check-act cycle, is used to design or adapt a process for a new product or service and to improve the quality of an existing service.

Phase I of IPM focuses on identifying and defining (1) the product or service to be delivered, (2) the end-user customer receiving the product or service, (3) the customer's specific needs, and (4) Liberty Bank's requirements of its internal and external suppliers to the process. Phase II involves mapping the production or service delivery process and assessing the quality of the output (product or service). In Phase III, PITs evaluate and improve the product or service process. IPM, which is depicted in Figure 21-1, also incorporates IDEAS where appropriate.

IPM provides for involvement of all process participants, and some of the steps require the participation of customers and suppliers. As illustrated in Figure 21-1, IPM is a continuous improvement cycle, requiring constant efforts to monitor, correct, and prevent process upsets, to benchmark similar processes at high-performing firms, and to explore new ways to improve the process.

In determining process improvement candidates, Liberty Bank considers several factors. As noted above, a highly efficient process may be a candidate for simplification. We also measure the quantity of paper consumed and produced by the process, continuously determining whether the physical outputs are really necessary or if the information can be distributed electronically. Customer satisfaction (or dissatisfaction) at moments of truth may also point to process problems. Finally, we have an aggressive program of competitive comparisons (Item 2.2), using available process data to evaluate our processes and outputs against those of world-class performers.

5.2c   At Liberty Bank, service delivery is often monitored as it happens, whether electronically, by employees, or by "mystery shoppers," so that performance data are available quickly. Other performance data are obtained from customers via surveys, focus groups, and direct contact with account representatives. We analyze these data according to the methods described in Item 2.3.

5.2d   Process improvements developed through IPM are integrated into daily operations and processes by revising process documentation and informing all other process participants and support functions of their use. Because the

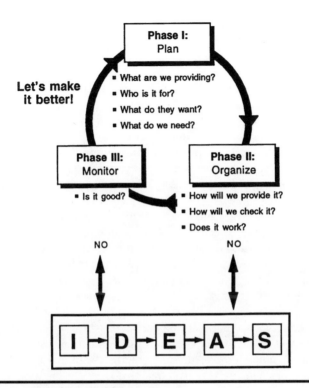

Figure 21-1 Integrated Process Management

process improvement teams comprise process participants, each team member is responsible for apprising his or her department or area of the changes and ensuring implementation. The process champion oversees this dissemination and integration and works with the process improvement team to monitor the effective use of the improvements and verify that the improvements have the desired effect. The Senior Management Quality Council and the staff of the Chief Quality Officer (CQO) monitor the actions taken by teams to ensure that the improvements and related process standards are effectively integrated. A corresponding effort is made to confirm that the related data incorporated into the QMICS system are accessible, understandable, and usable by all Bank units and any outside services or vendors.

5.3 Process Management—Business Processes and Support Services

5.3a Liberty Bank also uses IPM for quality control and improvement in its support services and business processes. Accordingly, all support processes have process champions that have the same responsibilities and authorities as the champions of product and service production and delivery processes.

When used for business and support processes, the planning phase of IPM allows for the translation of customer expectations into process requirements and service standards. In essence, the concept of "the customer" is driven across each step of the process, so that every employee in every function of the Bank recognizes that he is a customer and a supplier not only within a process, but also across processes. Again, effective and frequent communication at these process intersections (see Item 5.2b) is critical to cross-process linkages. Furthermore, process control standards, quality measures, and improvement goals for business processes and support services are fully integrated into the QMICS system. Applying the same guidelines, each support unit and business process is also subjected to rigorous internal and external audits.

Most business processes and support services have to be designed so that they are consistent with the needs of the external customer. For example, software modifications and updates are often motivated by increased demands from customers and regulatory agencies. Other internal requirements, such as timely expense-account reimbursement and accurate personnel file changes, do not affect the external customer, but are very important to employees, the *internal* customers.

As with processes serving external customers, support processes are measured with respect to internal moments of truth and are measured for the same service features—timeliness, accuracy, immediate transaction service, knowledge, courtesy, and accessibility. Internal customer satisfaction data are gathered quarterly; comprehensive employee surveys are usually conducted annually; and the benchmark Stanek surveys, which provide a valuable calibration of the internal surveys, are done every two years.

5.3b As with product and service production and delivery processes, all of Liberty Bank's support services have quality improvement goals. These are based on external benchmark data when available; otherwise, internal best practices are used. Again, process champions have process improvement "quotas" similar to those required for production and service delivery processes: where there is a demonstrated need for the improvement or when the process has been performing at or above standard for at least four quarters.

Improvements in support processes are sometimes motivated by improvements in comparable externally focused processes. For example, the improvements made in the process that gathers and reports 401(k) data to corporate clients were copied and used in the process responsible for reporting certain internal financial data. Other motivations include process information from suppliers and customers, improvements in computer and document technology, and advances in communications technologies and facilities.

6.0 QUALITY and OPERATIONAL RESULTS

6.1 Product and Service Quality Results

6.1a As detailed in Item 2.1, Liberty Bank's QMICS system tracks more than 160 quality and operational performance measures. Over the past three years, Liberty's performance for the most important service encounters with customers, presented in Figure 21-2, has shown steady improvement.

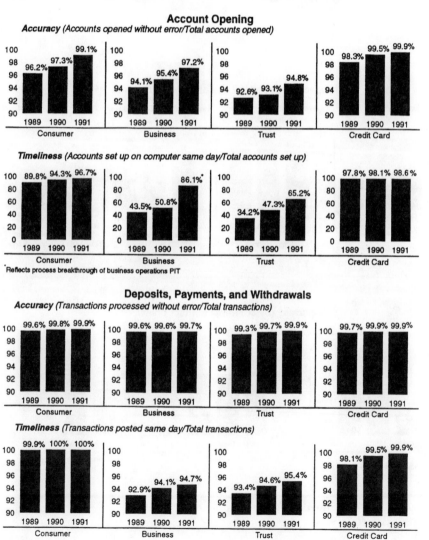

Figure 21-2 Performance for the Most Important Service Encounters with Customers

Inquiries (All operations)

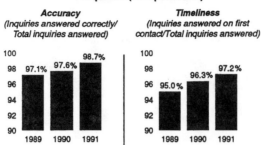

Accuracy
*(Inquiries answered correctly/
Total inquiries answered)*

Timeliness
*(Inquiries answered on first
contact/Total inquiries answered)*

Transaction Speed (by customer contact point)

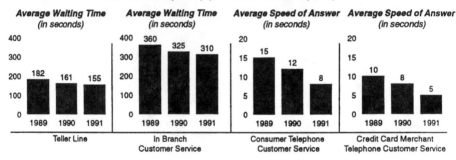

*Average Waiting Time
(in seconds)*

*Average Waiting Time
(in seconds)*

*Average Speed of Answer
(in seconds)*

*Average Speed of Answer
(in seconds)*

Teller Line

In Branch
Customer Service

Consumer Telephone
Customer Service

Credit Card Merchant
Telephone Customer Service

Courtesy (by customer contact point)

Interactions rated courteous/Total interactions monitored

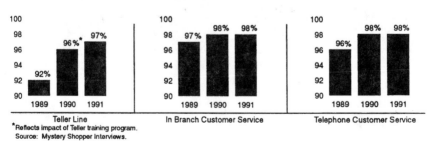

Teller Line
*Reflects impact of Teller training program.
Source: Mystery Shopper Interviews.

In Branch Customer Service

Telephone Customer Service

Figure 21-2 (continued) Performance for the Most Important Service Encounters with Customers

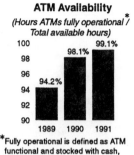

ATM Availability

(Hours ATMs fully operational /
Total available hours)

*Fully operational is defined as ATM
functional and stocked with cash,
receipts, and deposit/payment envelopes.

Other Business and Trust Measures

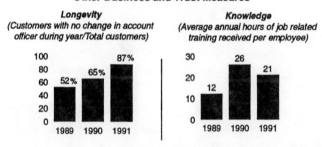

Longevity
*(Customers with no change in account
officer during year/Total customers)*

Knowledge
*(Average annual hours of job related
training received per employee)*

Figure 21-2 (continued) Performance for the Most Important Service Encounters with Customers

6.1b Regrettably, the financial services industry currently has relatively few reporting standards that would facilitate the comparison of performance across institutions. As a result, very limited "apples to apples" comparative data are available. The competitive comparisons presented in Table 21-4 were obtained from three primary sources:

▪ The "Top Banks" data were developed by a national consortium of 17 high-performing banks, including Liberty, with $1 billion to $5 billion in assets.

▪ The "Industry Average" data were compiled by the American Banker's Association.

▪ The "Top Baltimore Banks" data were developed through (1) a cooperative effort with our two largest competitors and (2) direct sampling by Liberty employees and mystery shoppers of our competitors' service levels.

Table 21-4 Competitive Comparisons

Service encounter/feature		Liberty Bank	Top banks	Top Baltimore banks	Industry average
Account Opening					
Consumers:	Accuracy	99.1%	99%	98.2%	98%
	Timeliness[1]	99.9%	99%	99%	98%
Business:	Accuracy	97.2%	NA	97.4%	NA
	Timeliness[1]	95.4%	NA	96.7%	NA
Trust:	Accuracy	94.8%	NA	92.7%	NA
	Timeliness[1]	83.4%	NA	67.5%	NA
Credit Cards:	Accuracy	99.9%	99%	NA	98%
	Timeliness[1]	99.9%	97%	NA	89%
Deposits, Payments, and Withdrawals					
Consumers:	Accuracy	99.9%	99.9%	99.8%	99.8%
	Timeliness	100%	100%	100%	100%
Business:	Accuracy	99.7%	99.7%	NA	NA
	Timeliness	94.7%	92.3%	NA	NA
Trust:	Accuracy	99.9%	99.9%	99.8%	NA
	Timeliness	95.4%	100%	100%	NA
Credit Cards:	Accuracy	99.9%	99.8%	NA	NA
	Timeliness	99.9%	99.9%	NA	99.9%
Inquiries					
Accuracy		98.7%	98.5%	96.2%	NA
Timeliness		97.2%	97.7%	93.7%	92%
Transaction Speeds					
Average	Teller Line	155 sec	NA	170 sec	240 sec
Waiting Times:	In-Branch Customer Service	310 sec	NA	330 sec	NA
Average Speed of Answer:	Consumer Telephone Customer Service	8 sec	10 sec	18 sec	20 sec
	Merchant Telephone Customer Service	5 sec	5 sec	NA	10 sec
Abandonment Rate					
Telephone Customer Service		3.2%	2.5%	3.4%	5.0%
Courtesy					
Teller Line		97%	97%	95%	NA
In-Branch Customer Service		98%	NA	96%	NA
Telephone Customer Service		98%	NA	98%	NA
ATM Availability		99.1%	99.4%	NA	97.5%

[1]Accounts set up in 48 hours.
NA = not available.

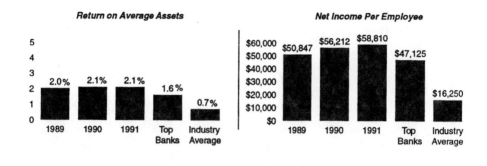

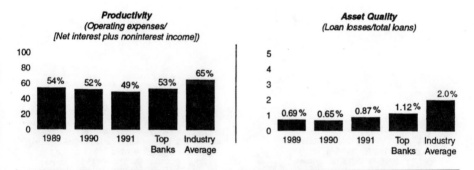

Figure 21-3 Profitability

6.2 Company Operational Results

6.2a, b There are only a limited number of company performance measures for which relevant comparative data are available. Comparative results for four company operational measures used in our industry are presented in Figure 21-3.

Our service levels have enabled us to become the primary bank for 75 percent of our customers, which is twice the proportion of primary customers as our competitors. By obtaining more business from each of our customers and improving the efficiency and effectiveness of our processes, we have been able to significantly outperform the industry and equal or outperform our peers with respect to the key measures presented in Figure 21-3.

6.3 Business Process and Support Service Results

6.3a, b As discussed in Item 5.3, all of Liberty Bank's support services track their performance in meeting agreed-to internal customer requirements. Of all the support areas, however, five have the largest impact on external customer service levels and employee satisfaction: Systems, Branch Operations, Mailroom, Human Resources, and Internal Audit. Results for the most important service features for each of these departments (as identified by their internal customers) are presented in Figures 21-4 to 21-8, together with comparative data where available.

Systems

Because of our high level of automation, our system performance has a significant impact on most customer interactions: from teller line transactions to telephone customer service to ATM availability. Our systems group has worked hard to continually improve the availability of our systems and the accuracy and timeliness of the hundreds of program modifications completed each year. Note: As discussed in Item 2.2, we have had only limited success in obtaining comparative and benchmark data for our business processes and support services.

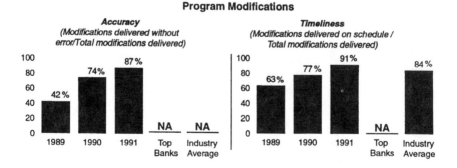

Figure 21-4 Systems

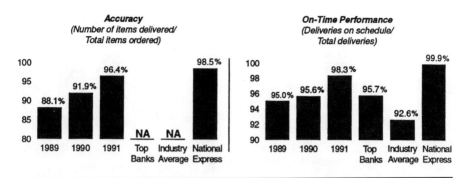

Figure 21-5   Courier Deliveries to Branches

Branch Operations: Courier Service

Since we purchase and receive all equipment and office supplies at the headquarters office, our extensive courier operation is critical to delivering these supplies to our 21 branches. If the couriers do not bring the needed supplies on time, branch customer service suffers. Our best-in-class benchmark for courier deliveries is National Express, the nation's largest and best-performing overnight package transportation service.

Mailroom

Our ability to meet our aggressive processing timeliness standards for correspondence is heavily dependent on our mailroom's accurate and prompt delivery of correspondence to the appropriate bank departments.

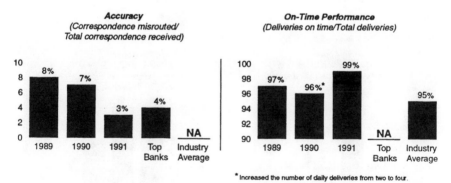

Figure 21-6   Correspondence Delivery

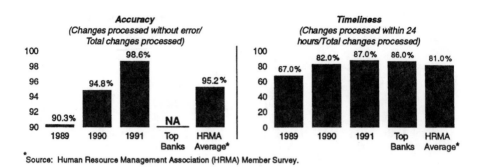

*Source: Human Resource Management Association (HRMA) Member Survey.

Figure 21-7   Employee Database Changes

Human Resources

Our employee database supports a variety of functions, including payroll, benefits, notifications and training eligibility. To better serve its customers—our employees—our human resources area has worked aggressively to improve the speed and accuracy of changes to the database (e.g., promotions, transfers, name and status changes).

Internal Audit

As detailed in Item 5.5, our Internal Audit area performs periodic financial and operational reviews of each area and branch within the bank. It has two customers, the management of the operating area and the Audit Committee of Liberty's Board of Directors. The expectations of these two groups are somewhat different.

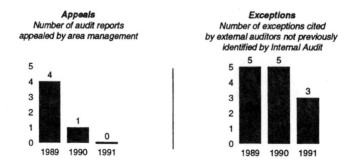

Figure 21-8   Audit Report Appeals and Exceptions

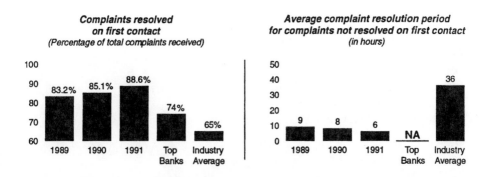

Figure 21-9 Complaint Handling

■ The management of each operating area wants Internal Audit to serve as a consultative partner, helping them to assess and improve their performance. Rather than the traditional adversarial relationship with audit, operational management feels Audit should (1) work with them to understand the root causes of performance issues, and (2) assist them in developing plans to address deficient areas. If the partnership is working properly, operational management feels that they should agree with the Board and not have to appeal audit reports.

■ The Audit Committee of the Board expects Internal Audit to be sufficiently thorough and rigorous that the external auditors do not find any "surprises" during their annual review.

Complaint Handling

While complaints are handled on a decentralized basis, as discussed in Item 7.1, Liberty has established aggressive Bankwide standards for complaint handling (Figure 21-9).

7.0 CUSTOMER FOCUS and SATISFACTION

7.1 Customer Relationship Management

7.1a Liberty Bank gathers a wide range of internal and external information on customer needs, wants, and expectations. Because Liberty Bank is a full-service bank offering multiple products and services to consumer, business, trust, and credit card customers, surveys and other customer and market analyses are structured to obtain information as specific to a product or service as possible. To ensure that we obtain objective, unbiased information, our research is conducted on a "blind" basis; respondents are unaware that we are sponsoring the research. To further ensure objectivity and validity, much of this research is conducted by independent market research firms. The research firm consults with us to ensure that the sample sizes are statistically valid and include the appropriate mix of primary and non-primary customers, as well as both lost customers and non-customers.

All of the individuals contacted are asked to:

■ Rank the relative importance to them of various existing or planned products, product features, and service features

■ Name, in open-ended questions, products and services that they would like to see that are not currently offered by their bank but that may be offered by other institutions.

■ Cite the reasons they prefer their primary bank to others they may use, and what the other bank or banks would have to do to win them as primary customers

Our main customer survey is conducted quarterly; others are conducted twice a year, annually, or biennially. Surveys to the same product base remain comparable over time so that trends can be observed. Where changes in satisfaction levels cannot be explained by analyses of internal measurements, we conduct in-depth focus groups with customers to develop a detailed understanding of why satisfaction has changed.

We also survey 1000 consumer and 250 business non-customers every year to determine their satisfaction with their banks and the reasons they selected their banks. All external surveys are coordinated by the Marketing Department to ensure consistency of wording and approach.

Over the years, our research has shown that eight service features are paramount in the minds of all customers, whether consumer, business, trust, or credit card. These features—timeliness, accuracy, immediate transaction service, knowledge, courtesy, convenience, accessibility, and longevity of the account officer/team—are tracked for all products and services (to the degree that they are relevant) on a moment-of-truth basis. We have developed standard (minimum acceptable performance) and high-level (stretch) targets for each of these features. For example, we measure our performance in handling incoming telephone calls to our Customer Information Center (CIC) for two of these factors *even before a telephone representative picks up the phone:* speed is a measure of timeliness, and call abandonment is an inverse measure of accessibility. In addition, using quarterly reports from our local telephone company, we evaluate trends in the number of calls that are "blocked" because all of our lines are busy.

We measure each of these factors for all customer groups and customer-contact processes. In the case of corporate disbursement services, for example, we have set a standard target of answering the phone within 10 seconds, and a high-level target of 5 seconds; because credit card authorizations can be performed electronically, we have set more aggressive targets of two rings and one ring, respectively. By tracking these factors, and interviewing customers about their satisfaction with our performance on them, we are able to validate and revise our targets periodically to ensure that we are delivering the levels of service required to delight and retain our customers.

Most of our service standards and policies are communicated to our customers. For example, our policy is to have deposits made from funds held in Maryland banks available within 24 hours. Accordingly, all deposit receipts from a branch of an ATM have printed on them the date that the funds will be available; branch personnel also tell depositors this information. Other policy information is printed clearly on statements.

Although Liberty Bank prides itself on being a service quality leader, there are factors affecting our products and services that are simply beyond our control, and many of these have the potential to cause customer dissatisfaction. One obvious example is the recent downward trend in interest rates. While this trend is beneficial to customers in that it lowers loan interest rates, it also reduces the interest they receive on deposits at Liberty Bank. We therefore make certain that every move of interest rates (including mortgage rates) is preceded by a notice to customers, whether in the monthly statement, on ATM screens, or in the automated call-answering system.

We also track the difference between the interest we pay and the minimum interest payable in U.S. Savings Bonds. Every six months, if the new savings bond interest rate is more attractive than our term deposit rates, we notify holders of expiring long-term (3- to 5-year) certificates of deposit that they may be better served by purchasing bonds and offer to make the transaction (by electronic transfer, if appropriate). Since we instituted this policy, the number of dissatisfied customers citing interest rates as a reason for their dissatisfaction has declined from 24 percent to 7 percent.

In our ongoing efforts to increase customer satisfaction, a QIT from the Customer Information Center (CIC) used the IDEAS process to perform an in-depth analysis of incoming customer calls to determine if some of those calls could have been prevented. During June and July 1990, the CIC customer-service representatives used check-sheets to classify each call they received based on the reason for the call. The QIT then compiled the data and developed a Pareto chart of call frequency by reason. To their surprise, they found that two of the top five reasons for the calls—status checks on inquiries and questions about the wording of Bank correspondence—were actually caused by the Bank!

Based on this analysis, the QIT recommended and the Bank implemented a number of specific actions to reduce the number of times we caused our customers to call us. Among these were customer panels to assess the clarity of Bank communications *before* they are sent out and the adoption of a Bankwide policy for handling customer inquiries. Specifically, anyone receiving a customer inquiry that cannot be answered on the first contact will:

■ Work out a mutually acceptable time to get back to the customer

■ Enter the inquiry on the CIC inquiry system

■ Notify the appropriate department or area

■ Check the customer's file in the CIC to verify that the Bank has kept its promise to respond

Since implementing these measures in late 1990, the number of "avoidable" calls received at the CIC has declined from 15 percent of total calls to less than 6 percent. This reduction has in turn lessened the need to hire additional customer-service representatives and has enabled us to provide better service on the calls we do receive. The program's success has also motivated the CIC team to work on further reducing avoidable calls, and several other Bank departments have begun to analyze their inbound calls to see if they can be prevented.

Our QMICS online information system has proved invaluable in supporting our quality improvement efforts. All customer-contact personnel are trained in all facets of the system before they assume their roles. Terminals throughout the Bank allow virtually all customer-contact employees to call up a customer's relationships on the screen and, in most cases, to answer the customer's question without a callback. Access to a customer's relationship history is also provided at each teller window, enabling branch customers to receive answers to most account questions without talking to a second Bank employee. Because all accounts are linked in the Customer Information File (CIF), all account inquiries are tracked for later analysis.

7.1b Customer service is integrated with overall management in a number of ways.

■ We work to create a high level of awareness of overall customer requirements through internal campaigns and constant reinforcement and train employees in how to identify and meet individual customer needs.

■ We provide two toll-free telephone number (1 800 MCVISA1 for credit card customers and 1 800 LIBERTY for all others) to allow customers easy access to the Bank. We include these and other telephone numbers on all statements and other customer communications so that customers know how to get information or assistance.

■ Our ATM network is among the most sophisticated in the country. In addition to the usual transactions (deposits, withdrawals, transfers, and balance inquiries), our customers can also obtain recent histories of their accounts. Screen displays show their five most recent deposits, withdrawals, and paid checks; the status of any check written since the last statement; total deposits, payments, and interest since the last statement; and the current value of any CDs they may hold.

Phones at each ATM provide direct access to our 24-hour Customer Information Center (CIC) so that any question can be answered immediately. Customers are very satisfied with our ATM system: 86 percent say that it is the best system they have used, and nearly half use the ATMs as their primary method of making Bank transactions.

■ We also participate in the PRODIGY™ network, so that customers can access similar information on their accounts from their home computers; corporate customers receive terminals as part of their account relationship. Once customers log on and enter their passwords, they can access and analyze a broad range of account data, as well as print accessed information on their home printers. Although this program is less than 1 year old, 8 percent of our customers are currently using it, and our quarterly survey indicates that another 10 percent plan to begin doing so in the next 6 months.

■ Our CIC has been upgraded to provide 24-hour access for customers seeking assistance. Through the CIC, customers have access to information on all products and services in the relationship of a single call, thus simplifying the inquiry process.

■ Corporate and trust accounts have dedicated officers and teams who are always available to respond to customer inquiries.

7.1c In addition to the various customer satisfaction surveys that Liberty Bank conducts, we routinely survey customers who have recently opened, changed, or closed an account or purchased a brokerage or credit product. These surveys cover how the customer was served in the transaction and how effectively the Bank's product or service met the customer's needs. We also conduct routine post-inquiry or -complaint follow-up calls and surveys to determine if the customer was satisfied with our response or dissatisfied with any aspect of it.

7.1d Our Code of Conduct is derived from customer requirements and includes a list of well-defined, objectively measurable standards. The premise of the Code is clear: *All customers are important, no matter how much business they do with us.* The employee and supervisor both sign the Code, and each unit manager is responsible for monitoring employees' adherence. The specifics of the Code are as follows:

1. Always be eager to serve customers.

2. Apologize for any delay in service regardless of who created the problem.

3. Thank the customer at the completion of every transaction.

The Code of Conduct and customer-service standards are reinforced throughout the Bank, particularly among customer-contact employees. Department and unit managers monitor at least ten customer-service interactions (by phone and in person) per month. Simulated customer calls are placed randomly to test conformance to standards, and "mystery shopper" visits occur routinely in each branch. Employees' attitudes and compliance with the Code of Conduct are then reviewed with employees monthly; these reviews in turn are used in periodic performance appraisals. The reviews are meant not to criticize but rather to coach employees on how to deliver outstanding customer service and to reaffirm the Bank's standard on such matters as proper telephone use and how to treat customers. To ensure that the reviews are constructive and positive, we have trained all of our managers and supervisors in the basics of effective performance appraisals.

The continuous improvement of service standards is one of the key responsibilities of all quality teams. Accordingly, our performance relative to our service standards is reviewed monthly through QMICS reports and quality reviews.

7.1e Complaints are directly managed by each of the four major product areas: consumer, business, trust, and credit card. Each area relies on the Bank's computerized Complaint Tracking System (CTS) to ensure that all complaints (formal and informal) are resolved and to facilitate the collection and analysis of complaint data.

Consumer complaints are received by telephone at the Customer Information Center (CIC), by telephone at the branches, or in person at teller windows. To facilitate first-contact resolution, we permit customer-contact personnel to settle balance and deposit disputes in the customer's favor, up to $100 per occurrence. This policy helped us to resolve 89 percent of all complaints received in 1991 on the first contact.

In certain cases, such as information discrepancies or an employee's incomplete product or service knowledge, the Bank has a formal escalation procedure. For complaints not made directly to the CIC, customers are told that a CIC representative will join the conversation. In most cases, the CIC's centralized capability allows for immediate action; when this is impossible, the CIC representative assumes ownership of the call, follows the four steps outlined in Item 7.1a for inquiries, and then notifies the branch of the resolution. For their part, branch personnel record in the CTS the fact that they had to pass the complaint on to the CIC.

The Business and Trust Account Officers have primary responsibility for resolving complaints from their customers. This is generally done directly with the customer and the head of the appropriate department or unit. All complaints are given the highest priority, reflecting the belief that if the customer has taken the time to complain, the relationship can be improved. The Credit Card unit has its own process for receiving and handling complaints, since many of the sources of dissatisfaction are unique to its position as a direct-marketed product.

The CTS provides the primary means for ensuring that complaints are resolved promptly by tracking the time needed to resolve each complaint. Daily complaint aging reports are produced that indicate which complaints are approaching the commitment date agreed to with the customer or the standard resolution date as defined by type of problem. Complaints that have not been resolved within 24 hours are investigated to determine the cause and whether any additional resources or assistance is required. The CTS also permits the aggregation of complaints across all customer groups.

With CTS, 98 percent of complaints are now resolved within 24 hours, as opposed to 23 percent four years ago, and 96 percent of customers surveyed are satisfied with the resolution of their complaint, compared with 35 percent four years ago.

7.1f Liberty Bank recognizes that customer satisfaction is determined by how well the Bank meets customers' expectations at each moment of truth. Accordingly—and in sharp contrast with the approach taken by competitors—customer-contact is not an "entry-level" job at Liberty Bank, but rather a career position. Bank employees are not eligible for promotion to customer-contact positions until they have had at least three months of satisfactory "back-office" experience, where the employee learns about the Bank's full line of products and services, as well as the banking industry in general.

Nonsupervisory customer-contact personnel in our Customer Information Center (CIC) handle telephone and mail inquiries from all customer groups and receive six weeks of CIC training before they are certified to answer a phone or respond to written correspondence on their own. Again, CIC employees must have demonstrated satisfactory performance and knowledge of the Bank's products and services before they are eligible for these positions.

The Bank recruits individuals for customer-contact positions with the skills, knowledge, and psychological profile to excel in their jobs. We then provide them with comprehensive training in Commercial Loans to Businesses; Managing the Quality of Customer Service; Remember Me, Your Customer; Remember Me, Your Supervisory Group; and Creating Satisfied Customers.

Because Liberty Bank has a policy of "staffing up" during peak business hours, many of the branch customer-contact positions are occupied by part-time employees who have satisfied the same three-month experience requirement. Most of these part-time employees are rejoining the work force after spending time at home with small children. Liberty Bank specifically targets this segment of the work force for its strong work ethic and demonstrated patience.

Business and trust account officers are expected to understand the customer's perspective and to use good judgment to negotiate mutually rewarding agreements. Likewise, all employees, particularly customer-contact employees, are required to take the actions necessary—within the confines of regulatory guidelines and Bank policy—to resolve any customer problem. This philosophy is reinforced through the Bank's training and awareness programs and by recognizing and rewarding those individuals who have gone to great lengths to satisfy customers.

Liberty Bank determines the attitude and morale of customer-contact employees primarily through its annual Employee Survey and maintains a number of recognition programs at both the branch/department and Bankwide level. Performance appraisal of these employees includes comments from customers and peers as well as the input from the monthly Code of Conduct reviews (see Item 7.1d). At each performance review, the employee and the supervisor discuss the employee's strengths and areas for improvement and develop personal improvement goals and plans.

Customer-contact employees who consistently delight their customers (as shown by high customer satisfaction scores) are eligible for promotion to Customer Account Manager, the first exempt level in the customer-contact career path. Customer Account Managers work with high-worth consumers and business or trust customers and are encouraged to develop their own contacts and accounts. On the other hand, Liberty Bank has also broken the required management track for high-performing employees by providing appropriate levels of respect and compensation for long-term employees who wish to remain in their customer-contact positions. These employees often become "mentors" or sponsors for younger staff members (and their supervisors often turn to them for advice).

Account officers are judged on the basis of their ability to manage relationships. Fifteen percent of each account officer's salary is considered "at risk," with full receipt of that portion requiring customer satisfaction ratings of 90 percent.

7.1g Liberty Bank evaluates its customer relationship management practices by soliciting customer comments and then by examining employee comments and concerns, as expressed in surveys, during appraisals, or during interviews with PITs. Customer surveys ask for ratings on all aspects of relationship management, including personnel, procedures, and performance. Employees are asked to describe circumstances in which they felt that their own performance or responses were insufficient or potentially unsatisfactory, to note ways in which the interaction with the customer could have been improved, and to identify policies, procedures, and practices that, in their opinion, hinder the delivery of superior customer service.

Liberty Bank also measures and tracks the performance of the CIC and CTS, the effectiveness of its hiring and training programs for customer-contact personnel, the morale and retention rates of these personnel, and customer satisfaction with its various practices. Analysis of these data enables the Bank to determine areas for improvement.

7.4 Customer Satisfaction Results

7.4a Liberty Bank tracks customer satisfaction with its service performance on a moment-of-truth basis across eight major service factors (see Item 7.1) and aggregates customer satisfaction data by customer group, specifically consumer, business, trust, and credit card. Through the many efforts described throughout this application, we have achieved ever-increasing levels of customer satisfaction over the past five years in key areas.

As described in Item 7.3, all of our customer satisfaction instruments have a seven-point scale. In Figure 21-10, "Percentage Satisfied" represents the percentage of respondents giving a rating of 5, 6, or 7 on this scale. The data are for those customers who have had recent transactions and who named Liberty Bank as their primary bank.

7.4b Liberty Bank currently has no claims or judgments against it, nor has it had any major claims or liability judgments in the past seven years. Due to the nature of Liberty Bank's business, it is not subject to recalls or returns. Liberty Bank has always been judged to be in compliance with state, local, and federal regulations. No consumer complaints have been made to bank regulating agencies for six years, and we know of no major or minor issues of concern.

The primary indicators of customer dissatisfaction are complaints and closed accounts for reasons other than geography. For each customer segment we track complaints and customer perceptions of Liberty Bank's response to the complaints and evaluate closed accounts. These are presented in Figure 21-11; again, "Percentage Satisfied" represents the percentage of respondents giving a rating of 5 (Satisfied), 6 (Very Satisfied), or 7 (Delighted) on Liberty Bank's seven-point scale (see top of Figure 21-10).

7.5 Customer Satisfaction Comparison

7.5a As described in Item 7.3, our competitive comparisons are not as detailed as our analysis of internal operations. For example, our comparative research does not cover as many different moments of truth for our competitors' customers as it does for our own customers. The comparative data we do collect, however, clearly demonstrate that our performance compares favor-

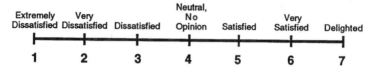

Customer Satisfaction Scale

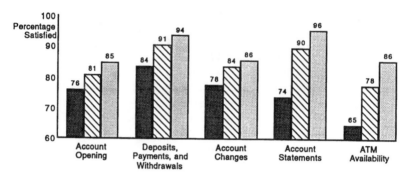

Customer Satisfaction—Consumer Accounts
(by moment of truth)

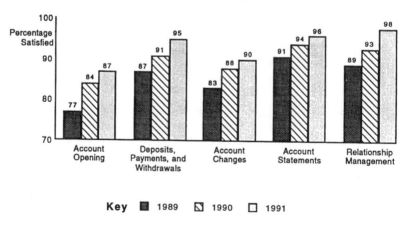

Customer Satisfaction—Business Accounts
(by moment of truth)

Key ■ 1989 ◨ 1990 ☐ 1991

Figure 21-10 Customer Satisfaction

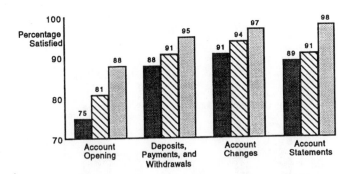

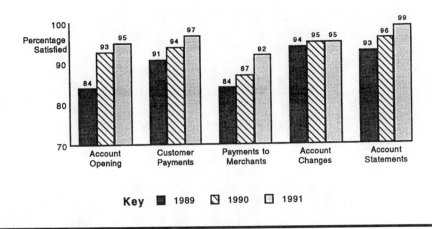

Figure 21-10 (continued) Customer Satisfaction

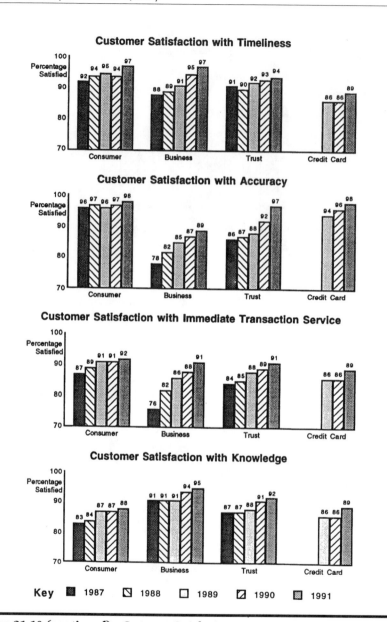

Figure 21-10 (continued)   Customer Satisfaction

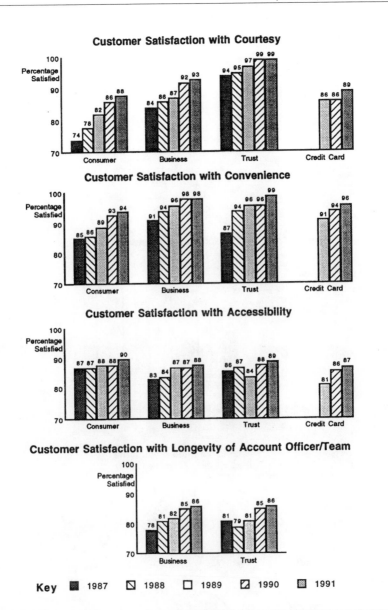

Figure 21-10 (continued) Customer Satisfaction

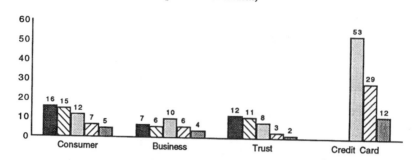

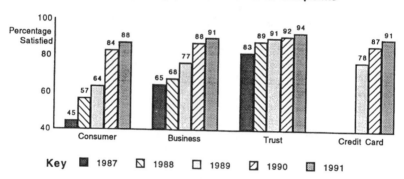

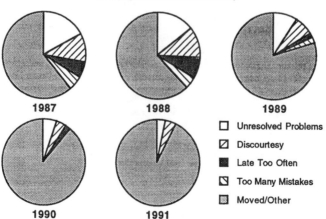

Figure 21-11 Customer Dissatisfaction

Table 21-5 Customer Satisfaction Comparison—Primary Customers (percentage satisfied by moment of truth)

Moment of truth	Liberty	Top Baltimore area competitors	Baltimore market average
Account opening	85	79	64
Deposits, payments, withdrawals	94	83	77
Account changes	86	86	72
Account statements	96	92	68
ATM availability	86	80	75
Relationship management	98	95	67
General inquiries	87	87	80

ably with the best in the industry (Table 21-5 and Figure 21-12). As before, "Percentage Satisfied" represents the percentage of respondents giving a rating of 5 (Satisfied), 6 (Very Satisfied), or 7 (Delighted) on Liberty Bank's seven-point scale.

7.5b A review of the switching patterns of consumer and business customers in the Baltimore SMSA indicates that 8 percent of customers switch from one Baltimore bank to another every year. Liberty Bank is one of the two major players in the market experiencing a net gain over the past three years. While 6 percent of our customers left in that period, 9 percent of the customers of competitor banks opened accounts with us, resulting in a net market share gain of 3 percentage points. Surveys of new customers indicate that 62 percent of our new customers chose Liberty Bank over other nearby banks because of our reputation for fair prices and high service quality.

Another indicator of market gains in the Baltimore SMSA is related to mortgage refinancing. Liberty Bank sells most of its mortgages on the secondary market, so it is not at risk from losing mortgagees to refinancing from other institutions. On the other hand, the Bank does keep track of how many of the "refis" it originates are to replace mortgages originally obtained through other institutions. In 1991, when the number of refis in the region increased 244 percent, the Bank found that half of its refi customers obtained their original mortgages from other Baltimore-area institutions. They cited Liberty Bank's interest rates and mortgage application process as their primary reasons. It stands to reason, of course, that the other half of Liberty Bank's refi customers obtained their original loans from Liberty Bank itself. Surveys have shown that this "repeat rate" is unusually high, with other area banks averaging a repeat rate of less than 20 percent.

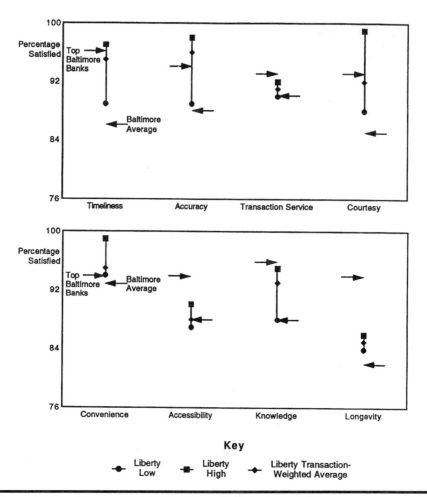

Figure 21-12   Customer Satisfaction Comparison by Service Feature

Liberty Bank has also made significant gains in the credit card market, where consumers nationwide are free to obtain their VISA or MasterCard accounts anywhere. Because of the Bank's aggressive marketing and attractive interest rates, the number of credit card accounts originating outside the state has grown 64 percent since 1988; according to surveys, more than three quarters of these new accounts replaced consumers' existing accounts.

7.5c   Liberty Bank has shown steady and significant gains in Baltimore-area market share for all types of accounts (Figure 21-13). Through 1990, the Bank improved its position with respect to its primary competitors. In 1991, we

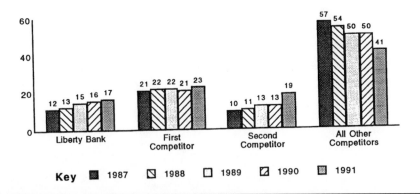

Figure 21-13 Market Share Comparisons (percentage of all accounts)

appeared to lose ground to one of our competitors, but that bank's increase in share was the result of its merger with another Maryland institution that had a significant Baltimore-area presence. (Liberty Bank chose not to pursue the smaller bank because management judged its commercial real-estate holdings to be of poor quality.)

INDEX